EMERGING TRENDS IN
BIOGAS TECHNOLOGY

EMERGING TRENDS IN

BIOGAS TECHNOLOGY

A. SAJIDAS

First edition

Published by :
Dr. A. SAJIDAS
Managing Director
BIOTECH INDIA Renewable Energy Pvt. Ltd.
BIOTECH TOWERS, MP Appan Road, Vazhuthacadu
Thycadu Post, Trivandrum, Kerala, India
www.biotech-india.org

First print 2024

ISBN 978-93-5912-933-4 Hardback
ISBN 978-93-5980-741-6 Paperback
ISBN 978-93-5917-052-7 Digital download and online
ISBN 978-93-5996-703-5 E-book reader

Cover photo :
A. SAJIDAS,
Water jacket model portable biogas plant with stove

Cover Design, Typesetting, Layout & Printed at
Orange Printers Private Limited
Thiruvananthapuram 695 001
www.orangexpress.com

Photos edited by Vipin R. BIOTECH

Printed in India

Preface

The world is facing numerous environmental challenges, including climate change, air pollution, and waste management problems. Biogas technology is emerging as a promising solution to address these issues while also providing a sustainable energy source. Biogas is a renewable energy source that is produced by the anaerobic digestion of organic materials such as agricultural waste, municipal solid waste, and sewage sludge.

Industrial and commercial biogas plants have been operating efficiently for many years in developed countries. India and China have been leading the way in building biogas plants for both domestic and institutional use. With the discovery that all kinds of organic wastes can be used to replace cattle dung as the main raw material for the operation, there is huge potential for domestic and institutional biogas plants. Unlike the models used in the early stages of biogas projects, portable biogas plants have been developed in various models that can be used like any other household appliance for the generation of biogas from kitchen leftovers.

This book, "Emerging Trends in Biogas Technology," provides a comprehensive overview of the current state of biogas technology and its potential for future development. The book covers a wide range of topics related to biogas technology, including the biology of anaerobic digestion, the design and operation of biogas systems, the economics of biogas production, and the environmental benefits of using biogas as a fuel.

This book has been written to cover the basic information needed to understand the potential of biogas plants. At this time, when the demand for biogas project is increasing day by day, many entrepreneurs want to enter this field. The main factor that influenced the writing of this book was to provide the basic information needed by such people. Basic knowledge is highly essential for designing various models of biogas plans. Studies show that unscientifically designed biogas plants cause serious air pollution. The content of this book is to help readers understand more about various types of domestic and Institutional biogas plants developed under the leadership of BIOTECH INDIA, which is working towards the development of biogas projects. The book also includes a special chapter on nano biogas digesters that can be used in both research, awareness programmes and domestic applications.

Although there are many books available for research students and teachers, this book is styled in such a manner that those with only basic knowledge can quickly understand biogas technology. Overall, this book is intended to serve as a valuable resource for anyone interested in the potential of biogas technology to provide sustainable and environmentally friendly solutions to our energy and waste management challenges. There may be some omissions and errors. Your valuable comments and suggestions on this book are invited. It will help make the subsequent editions even better.

Forward

DR. ATMA RAM SHUKLA,
President, India Biogas Association,
Former Advisor, Bio-energy Division,
Ministry of New and Renewable Energy,
Government Of India

This Book entitled, 'Emerging Trends in Biogas Technology', written by Dr. A. Sajidas is one, which encompasses various aspects of biogas technology. The types of materials which can be digested, types of technology which can be used, microbial process involved, various stages of digestion involved inside the plant, different components of biogas plants has been incorporated in this book, in addition to other relevant features.

It had been written in such a simple and comprehensible manner that any common person can also understands and use it while designing or using the plant.

The experience of Dr. A. Sajidas over the years in this technology from the very beginning has been collected together and put in this book in such a manner that anybody who is designing and using the technology can be benefited. It talks of the types of biogas plants and different kind of its utilization like cooking, electricity generation and fuel for vehicles. How to purify the gas to make it vehicle fuel, how to generate electricity from the biogas has been narrated in a very simple manner.

The plants covered are mainly small and medium size biogas plants and more interestingly the nano biogas plants designed for training, research and actual usage purposes.

The portable biogas plants are the ones which can solve the problem of tackling green waste produced in different kitchens of the country, either in rural areas semi- urban and urban areas, Thus, the problem of organic waste treatment is taken care off together with tackling bad smell floating around the houses in these areas. While treating the waste, it will also generate energy beneficial for utilisation as cooking fuel by the same family.

The bonus of organic manure which is generated as slurry from these plants can be used for vegetable cultivation in the same house or in the gardens of the locality. Thus, organic fertilizer which also comes as an output other than the gas has also been suitably covered in the book.

So this book, once it is in the hands of people, all across various stakeholders, are going to be benefited, I wish very well to Dr. Sajidas and the readers who are going to read the book and use it for various purposes, either for designing the plant or if they already have the plant then for operating it or if they are having any difficulties and plant is not working, the tips have been given to solve these problems and take appropriate corrective actions.

Dr. Atma Ram Shukla

Acknowledgments

I would like to express my deepest gratitude to Shri. Atma Ram Shukla (Senior Adviser Retired, Bio-Energy Division, Ministry of New and Renewable Energy, Government of India) who has written a touching forward to this book.

Sincere acknowledgements are due to those who helped me in writing this book. Proofreading and translation were done by Shri. D. Cyrus is specially remembered. Special mention is also made to Sri. Girish kumar, Mrs. Kabani. KS and Tejal Tilak Bakkar who helped in the translation.

I am also grateful to my colleagues and mentors who have provided valuable feedback and insights that have helped me improve my writing.

I would like to thank my family for their unwavering support and encouragement throughout this process. Their love and belief in me have been my constant source of strength.

I also wish to thanks all the staff members of BIOTECH INDIA Renewable Energy who provided the necessary support throughout the writing of this book. Lastly,

I would like to thank my readers for their interest in my work. I hope this book will inspire and inform, and I am grateful for the opportunity to share my ideas with you.

About the Author

Dr.A.Sajidas has the unique experience of more than 38 Years in designing, execution and the development of various Domestic biogas projects, Waste to Electricity Biogas projects, Organic waste management , Organic farming, Sustainable agriculture development, Energy conservation activities etc... Presently he is working as the International Consultant as well as the Managing Director, BIOTECH INDIA Renewable Energy with headquarters at Trivandrum, the capital city of Kerala state, India.

He has to his credit several inventions related with the generation of green energy, organic waste management etc. which are useful to the rural public as well as the common people belonging to different segments of the community.

He has published a good number of articles relating to green energy and waste management in various dailies and periodicals including "Akshaya Urja" an official publication of the Ministry of New and Renewable Energy, (MNRE) Government of India.

Dr. Sajidas is recipient of several National and International awards including the INTERNATIONAL ASHDEN AWARD-2007 from the UK, for generation of renewable energy through decentralized waste treatment. This prestigious award was Presented to him by Al Gore, former American Vice President.

He worked as an International Biogas Consultant in The Ministry of Agriculture and Irrigation, Republic of Yemen, under the project of "Biogas Digesters- an Integrated Solution for poverty alleviation and climate change, piloting in Yemen" funded by the World Bank and the Japanese Social Development Fund of the Government of Japan.

He has also to his credit, the designing and development of a portable Domestic Bio Waste Treatment-Biogas plant for treating the food waste and organic waste water, by converting these into cooking gas. He has a unique R & D style of functioning wherein he and his team, work incessantly. These enabled him to come forward with several inventions. A man of ideas and initiative, Dr.A.Sajidas with his flagship BIOTECH INDIA, has laid a landmark in the World map.

He is also serving as a member of the Expert Committee for Approving New Biogas Plants in the Ministry of Non-Conventional Energy, Government of India.

He is constantly working to bring the biogas project to the common people. He has also designed many courses that are beneficial for those who want to enter biogas and waste management sector. In addition to the above activities Dr.A.Sajidas is a Technical Advisor of different reputed companies in India and abroad. More details about the author is available in the web site www.drsajidas.com

Contents

List of Pictures

Acronyms

AD	-	Anaerobic Digestion
BOD	-	Biochemical Oxygen Demand
CBG	-	Compressed Bio Gas
CER	-	Certified Emissions Reduction
CNG	-	Compressed Natural Gas
COD	-	Chemical Oxygen Demand
CN	-	Carbon to Nitrogen ratio
CDM	-	Clean Development Mechanism
Cum	-	Cubic Meter
FRP	-	Fibre-glass Reinforced Plastic
GI	-	Galvanized Iron
GS	-	Gold Standard
HDPE	-	High Density Polyethylene
HRT	-	Hydraulic Retention Time
Kg	-	Kilo gram
KVIC	-	Khadi and Village Industries Commission.
Ltrs	-	Liters
MNRE	-	Ministry of New and Renewable Energy
MSW	-	Municipal Solid Waste
NPBD	-	National Project for Biogas Development
NGO	-	Non-Government Organization
NBMMP	-	National Biogas and Manure Management Programme.
NIRD & PR	-	National Institute of Rural Development and Panchayathi Raj
PVC	-	Poly Vinyl Chloride
RCC	-	Reinforced Cement Concrete
R&D	-	Research and Development
RET	-	Renewable Energy Technologies
UASB	-	Up-flow Anaerobic Sludge Blanket
UN	-	United Nations
UNEP	-	United Nations Environment Programme
UNDP	-	United Nations Development Programme
VER	-	Voluntary Emissions Reductions
WHO	-	World Health Organization
WB	-	World Bank

Chapter 1
Biogas Technology

History of Biogas

It was in the middle of the 18th century that the presence of an inflammable gas was identified from the gas collected from sewage lines. This finding was the result of a detailed study done on the large amount of gases emitted from the sewers. The above discovery prompted scientists to conduct further research in this area. This news spread to different parts of the world very quickly. Realizing this, many people came forward to conduct research on this. This was followed by experiments and research to produce a combustible gas from bio-waste. As a result of continuous efforts, it was found that all types of organic waste could be treated to produce combustible gas It was also found that there were large fluctuations in the quantity and quality of the gas that can be generated from each type of organic waste.

Subsequent studies confirmed that bio-waste could be used to produce combustible gas for industrial, public and private use. Thus, the gas produced in this way was promising to be used as an alternative to traditional fuels. This gas came to be known as "Biogas" as it is generated from biomaterials.

Generation of Biogas

Through research and developmental studies done around the world, the potential for biogas production developed in a variety of ways. The research into biogas gained more importance with the shortage of conventional fuels.

The discovery that the gas produced from bio-wastes could be used as an alternative to conventional fuels such as petroleum, coal and natural gas led to a major breakthrough in this field. The current situation is such that the conventional fuel sources are declining day by day. At this point, the importance of an alternative fuel source is seen to increase tenfold.

It has also been found that the air pollution caused by the use of gas from bio-waste is very low. The gas from bio-wastes being a clean fuel is of great importance as the use of conventional fuels cause air pollution.

As a gas produced from a variety of organic wastes at a relatively low cost, other fuels can rarely be considered as an alternative to this gas.

Also, the availability of raw materials is not an issue for the production of this gas as bio-waste is easily available.

A large number of institutions that could not achieve the expected growth due to the unavailability of adequate fuels began to improve with the advent of biogas. It is also noteworthy that with the advent of biogas, developmental activities have gained momentum even in the rural areas.

With the discovery of biogas, tremendous changes have taken place in the living conditions of the people and also in the growth of different industries. A study conducted on the development of different types of biogas projects around the world shows that India has achieved substantial progress in the development of biogas in various sectors.

Biogas from Cattle dung

The production of biogas from cattle dung was discovered by the middle of the 19th century. As a result of incessant research, many designs of biogas plants were developed in different countries. India and China took many serious steps to make the biogas programme a people friendly programme.

Compared to many other countries in the world, India has given special importance to the biogas project and has carried out continuous research. Many research institutes, both big and small have contributed significantly to the development of this field.

In the beginning, biogas project was a scheme meant only for the farmers who owned cattle. Traditionally cattle breeders used to, store the daily cattle dung, dry it and burn it instead of firewood. The housewives who cooked with this dried dung were exposed to various diseases due to the huge quantity of soot and smoke generated during the burning of such dung. Of these, a majority was affected with respiratory and eye diseases.

The discovery that biogas could be produced from livestock dung led to a major revolution in the livestock breeding sector. It was a great blessing for the cattle farmers to be able to produce biogas for their own use from their own livestock. In addition, the use of slurry obtained after biogas production, which is a rich organic manure led to an increase in agricultural yields. Both of these are an excellent source of extra income for livestock farmers.

Numerous research and development activities have been carried out to design suitable biogas production plants for livestock breeders. Special research has been done in this field to make this project the simplest and one that can easily make anyone become a beneficiary. Considering the large possibilities of this project, governments in various countries have also implemented subsidy / incentive schemes for those setting up such biogas plants.

In the early stages, the project proceeded in such a way that cattle dung or Gober was used as the only raw material. For this reason, for a long time, the biogas project was only known as the 'Gober Gas Project'.

Chapter 1- Pic:1 Gober gas plant

Biogas from organic waste

Continuous researches have been carried out in different parts of the world for further development of this project so that it is not limited to only livestock farmers. Most of the research has been done to find out innovative methods to produce biogas from cattle dung and other organic matter. In addition to the Gober Gas Project, large-scale biogas plants have been developed in many countries for industrial biogas production. The invention of such gigantic plants also led to the elevation of biogas project to an industrial fuel production project.

Chapter 1- Pic:2 Institutional Biogas plant

Research into simplifying the biogas project and making it popular also progressed along with this.

It was in the beginning of the twentieth century that biogas technology developed in such a way that biogas could be used as a substitute for other cooking fuels by treating the everyday bio-waste in any household or establishment. Portable biogas plants were also developed during this period. In this regard, Biotech India Renewable Energy has played a pivotal role.

Chapter 1- Pic:3 Portable Biogas plant

Researches are being continued to find new technology to purify biogas to the same quality as that of natural gas and to use biogas to make vehicles run and to meet other energy requirements. This process is gaining popularity today.

As discussions on the declining availability of other fuels, climate change and global warming progressed, biogas related researches started gaining momentum. In addition to government research institutes, many private institutions and individuals also started doing research on their own. Experimental digesters suitable for such research work have been developed by BIOTECH. Experimental digesters of various sizes are now available to suit the needs of each individual. These are known as Nano digesters.

Chapter 1- Pic:4 Nano Digester

There are also very small biogas digesters suitable for households with only one or two occupants and very little bio-waste. With the advent of such Nano domestic biogas plants, biogas technology has become available to all sections of the society. As a result, every household can easily install biogas plants and treat kitchen waste to produce cooking gas. With the proliferation of biogas, the use of LPG can also be significantly reduced. The use of biogas would be a great relief to the beneficiaries in the current situation where LPG prices are rising day by day.

Chapter 1- Pic:5 Nano 500 Biogas plant

After continuous research, biogas technology has been developed into a scheme which can be used by all common people who have organic waste at their disposal. Many national and international organizations have already made significant strides in expanding biogas into a green energy project.

Biogas is now being used as an alternative fuel to meet the growing energy needs of both the private and public sector enterprises and households. The importance of this scheme is underscored by the fact that biogas plants are beginning to be utilized as a permanent solution to the ever increasing bio-waste treatment problem.

BIOTECH INDIA, a pioneer in the biogas sector developed portable biogas plants in the year 2000. This plant is capable of producing cooking gas from kitchen leftovers. This prefabricated and portable plant is a good substitute for other organic waste management systems. Different designs of portable floating dome model biogas plant have already been developed under the supervision and guidance of Dr. A. SAJI DAS, who is the founder of BIOTECH INDIA. More than 25 inventions have been made by him during the last 37 years of experience in biogas sector.

The New Indian Express

FRIDAY ■ AUGUST 18, 2000

Development of waste treatment plant claimed

Thiruvananthapuram, Aug 17: Biotech Centre for Development of Biogas Technology and other Non-Coventional Energy sources, a state Government award winner for energy conservation, has claimed to have developed a waste treatment plant producing biogas and biomanure from domestic waste. At a news conference here today, centre director A Saji Das said the live demonstration of the plant's working would be held at the centre's office premises at Vazhuthacaud M P Appan Road here from August 21 to August 22.ENS

Chapter 1- Pic:6 News report in New Indian Express Daily

BIOTECH INDIA thus plays a major role in the growth of biogas sector by helping in the transition from cow dung to daily available kitchen leftovers as the feed in biogas plants. There is remarkable scope for biogas technology which is easily accessible to the common man.

Chapter 1- Pic:7 Different model portable biogas plants

Chapter 2

Basic principles of Biogas generation

Let us examine the basic principles that apply to the production of biogas. Biogas is a mixture of various colorless and odorless gases. Biogas is produced through the fermentation of any degradable material (Substrate) with the help of different types of tiny microbes. A variety of bacteria, that belong to the group of microscopic organisms that cannot be seen with the naked eyes are involved in the production of biogas. Such microorganisms are known as anaerobic bacteria as they act in the absence of atmospheric air. There are several factors that influence this process, which is purely biological.

Biogas production takes place in a concealed chamber in the absence of atmospheric oxygen. The chambers which are prepared for biogas production should be airtight such that atmospheric air does not enter the chamber or the gas produced in the chamber does not leak out. This chamber is called the 'digester'. Digesters can be constructed in a variety of sizes and shapes. The type and the model of the digester can be determined based on the space available and the raw materials available for construction.

The biogas produced in the digesters is a mixture of different types of gases. These contain both wanted and unwanted gases in different amounts. Also, this mixture contains both combustible and non-combustible gases.

Biogas is a mixture of different gases; a major portion being methane (CH4) and carbon dioxide (CO2). It also contain traces of other gases like hydrogen (H2), Moisture (H2O), hydrogen sulphide (H2S) etc.

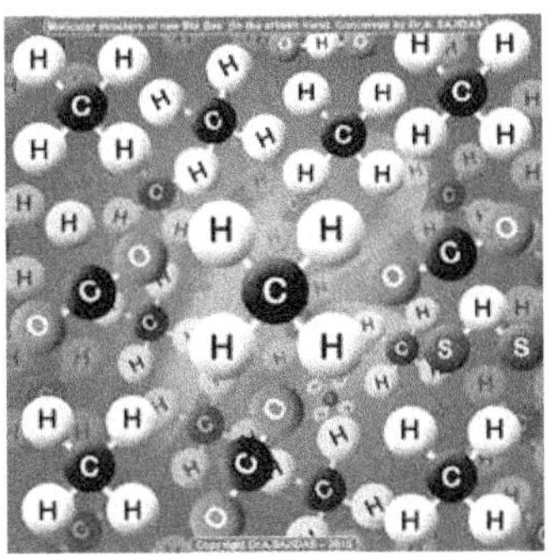

Chapter 2- Pic:1 Biogas molecular structure- an artistic view

The combustible component of biogas is Methane. As the amount of methane in biogas increases, so does the quality of the biogas. If the biogas contains high amounts of methane, the flame obtained during burning will be blue in colour. If the biogas is low in methane and high in carbon dioxide, the burning flame will have a yellowish blue colour.

The percentage of methane may vary depending upon the carbon-nitrogen ratio (C:N ratio) of feed materials (Substrate) and other factors influencing the fermentation process.

The composition of different gases in biogas varies depending upon the substrate utilized in the anaerobic digestion process. The biogas produced from old model Gober gas plants that are commonly used, contains about 60 percent methane and 40 percent carbon dioxide (60:40).

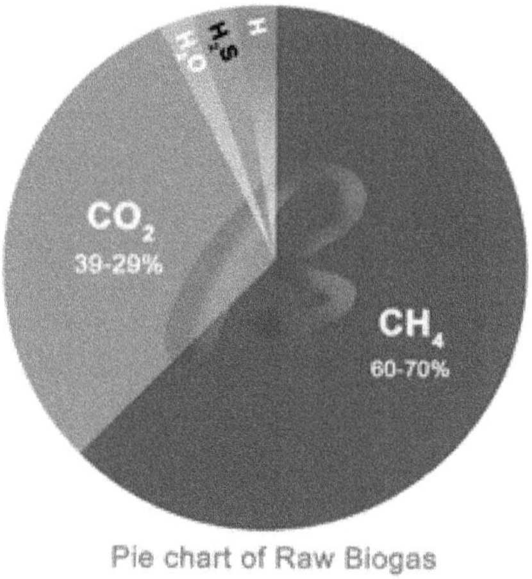

Pie Chart of Raw Biogas

Chapter 2- Pic:2 Raw Biogas chart an artistic view

The pH ratio of the raw materials used for biogas production also plays a crucial role in determining the level of methane in biogas. Higher the pH ratio higher will be the level of methane and lower the pH ratio lower will be the level of methane in biogas.

The biogas produced from biogas plants developed by using latest technologies contains 65%–75% of methane.

Biogas also contains water vapor. The amount of water vapor or moisture contained in biogas can fluctuate depending on the temperature inside the digester.

The water content in biogas must be removed before using it. Arrangements for removing the water content from biogas must be made along with the plant, in order to maintain the quality of biogas.

Another factor influencing biogas production is the atmospheric temperature. A certain level of atmospheric temperature is required for the efficient operation of microbes that affect biogas production. Biogas production is seen to decrease gradually when atmospheric temperature in the area where the biogas plant is installed goes below a certain level. Therefore, in order to avoid such a situation it is necessary to maintain the required temperature inside the digester.

Type of wastes (Substrate) that can be used to produce biogas

Technically, it can be said that biogas can be produced from all types of bio-wastes. But on a practical level, all kinds of bio-wastes can be used as feed, only in industrial biogas plants. In biogas digesters suitable for domestic and small establishments it is best to dispose of rapidly degrading bio-wastes only. If slowly degrading bio-wastes are deposited in small digesters, such waste will remain in the digester even after treatment. The bio-wastes remaining in the digesters will cause a reduction in performance of the digesters.

Industrial biogas plants operating with advanced configurations will have the facility to monitor all matters related to the functioning of the digester at all times and to make necessary amendments. These digesters will also be provided with the facility to automatically adjust the pH ratio inside the digester and also the carbon-nitrogen ratio of the raw materials.

The materials that are not suitable for biogas production are mainly non-biodegradable materials. Plastic, glass, metals and all other chemicals should be prevented from entering the biogas plant. Electronic and electrical waste materials should also be avoided. Non-biodegradable materials and chemicals act in two different ways once they reach inside of the digester. Of these, if non-biodegradable materials reach the digester they will accumulate inside the digester. As the amount of these materials increase, the digester activity is seen to slow down as the availability of suitable space for microorganisms to operate inside the digester decreases. If chemicals reach the digester, they will adversely affect the functioning of the microorganisms. If it exceeds a particular limit it can lead to the destruction of these microorganisms. Biogas production may reduce or completely stop as a result of depletion or complete extinction of microorganisms in the digester.

As domestic biogas plants are designed in the simplest way, it is best to use only rapidly degrading bio-waste as the raw material for biogas production. All degradable materials can be used for biogas production. But, it is best to use easily/ rapidly degrading organic waste as the raw material for biogas production.

Organic wastes including cooked food wastes, fruits and vegetable wastes, fish and meat wastes, water used for cleaning fish and meat, rice water, waste water containing food waste etc can also be used as raw materials for biogas production in domestic biogas plants.

Cow dung, chicken excreta, pig excreta, all kinds of animal and bird droppings and human excreta can be used for biogas production.

Apart from these waste materials, waste water from slaughter houses and food processing units can also be used for biogas production. Dried plant residues and particular plants like plantain can also be used as raw material in industrial biogas plants.

Industrial biogas plants have automated mechanisms for disposing bio-waste and controlling their discharge ratio. Industrial biogas plants can treat all types of organic matter, while inorganic and chemical wastes cannot be treated.

Ensuring that no inorganic or chemical contaminants are mixed with the bio-waste at the time of preparation of the bio-waste for disposal into the processing plant will help facilitate further operation of the plant.

Types of organic waste materials that should be avoided in domestic biogas plants

Slowly degrading vegetables, leaves of plants, grass, and egg shell having high fiber content should not be deposited in biogas plants. Wastes having high acid content such as sour curd, sour batter, decayed pickle, lemon peel are found to impair the plant function.

Types of organic waste materials that should be avoided in industrial biogas plants

In industrial biogas plants, as the raw materials used for gas production are collected from particular areas, it is unlikely that materials that could adversely affect biogas production will enter them.

Chapter 3

Biogas Generation Technology

The biogas generation technology is known as Biomethanation. Biomethanation technology is a centuries old technology. It was in the middle of the 18th century that the presence of a combustible gas was discovered from gases emitted from the sewers. It is a universally accepted and proven technology for Bio energy generation from organic wastes. It is effective in converting organic matter into biogas and bio-fertilizers as a result of the action of microorganisms that operate in the absence of atmospheric oxygen. It is very simple, user friendly and the carbon foot print of this technology is very less when compared to other energy generation technologies.

Through the adoption of biomethantion technology any degradable waste can be treated with the help of different types of anaerobic bacteria / microbes in a concealed chamber / digester. Biogas is a mixture of methane, carbon dioxide and traces of other gases. Treated biomaterials, that come out from the digester in liquid or semi liquid form can be used as a very good biomanure / organic manure.

Utilizing Biomethanation technology, all rapidly decomposing materials can be treated hygienically. The biogas generated through biomethanation can be utilized as a renewable source of energy for meeting all fuel requirements.

Stages of Biogas Production

For the generation of Biogas, the whole fermentation process of organic materials takes place in different stages. The three main stages are hydrolysis, acid formation and methane formation. Organic waste arriving at the digester is converted into soluble compounds by a specific group of microorganisms. This is called Hydrolysis. In the second stage, another group of microorganisms called Acid forming bacteria convert these soluble compounds into various acids. This is the acid formation stage. In the third stage, another group of microorganisms called methane generating bacteria convert these acids into biogas. This is the final stage of the biogas generation. All the activities in the above phases have to be continuous and simultaneous. The biogas thus produced will be collected in the biogas collector of the biogas plant.

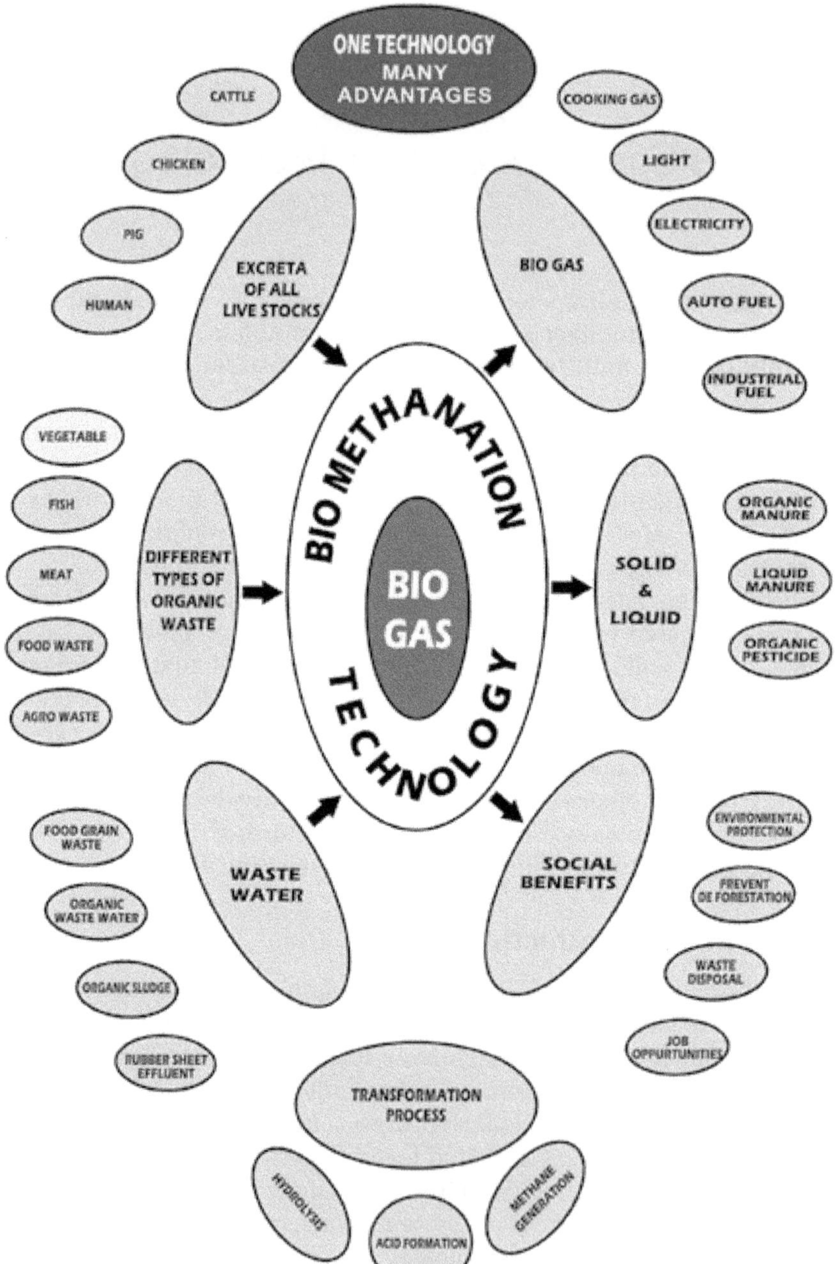

Chapter 3- Pic:1 Organic waste to biogas – process diagram

The generation of biogas from organic waste material takes place through the fermentation process, with the help of microorganisms. There are two types of anaerobic fermentations. They are wet fermentation and dry fermentation. Wet fermentation is commonly used in conventional biogas plants.

The digester of a conventional biogas plant is designed for continuous operation. The organic matter to be fermented is fed in a semi liquid form at one end of the digester and the fermented organic matter (substrate) is discharged at the other end.

For obtaining maximum performance, the fermentation process inside the digester has to be continued under certain specific conditions. When the fermentable materials are stored in a closed chamber, without any contact with atmospheric oxygen, will accelerate the growth of anaerobic bacteria. Anaerobic bacteria are tiny microbes that are not visible to the naked eye. Only under the microscope can one see the detailed structure of these anaerobic microbes.

The Acid forming group of bacteria is comparatively stronger and will survive in a wide range of conditions. Methane generation group of bacteria is very sensitive compared to the acid forming bacteria.

Depending upon the design of the plant, the biogas generated from the digester can either be stored in a gas collector constructed over the top of the digester or in a separate external gas collector(Balloon).

Chapter 3- Pic:2 External gas collector

There are a number of basic components needed to make a bio-waste treatment biogas plant efficient. Biogas production reaches its peak when all these components come together on the same level. For this it is essential to maintain equilibrium of the microorganisms operating at different stages of biogas production.

Biogas generation process (Fermentation)

Anaerobic digestion and Biogas production are biochemical processes occurring in different stages, during which different groups of microbes (bacteria) act on the organic materials.

The process by which organic matter gets decomposed as a result of the action of microorganisms is termed as fermentation. The fermentation process reaches its maximum level in conditions where there is no direct contact with the atmospheric air. Fermentation process is also known as digestion.

The gases obtained through this process also vary according to the quality of the raw materials used for the fermentation process. Special care should therefore be taken to select suitable raw materials to ensure high quality biogas production.

Certain parameters also have to be observed for the generation of biogas with sufficient quantity of methane. High quality biogas production can be achieved only if ideal conditions are available within the digester at each stage of the fermentation process. High quality fermentation can therefore be ensured by designing the digester properly and by using and selecting quality raw materials for biogas production.

Fermentation is an unavoidable process in various activities such as alcohol production and manufacturing of various types of medicines.

Digester

It is the most important part of a biogas plant. It is an airtight chamber. The organic materials which are fed into the plant undergo fermentation process inside the digester. There are different types of digesters available, based on the model of plant. In fixed dome model biogas plants, the digester and gas collector are constructed as a single chamber. In floating dome model plants, the digester and gas collector are fabricated separately and assembled together.

Chapter 3- Pic:3 Fixed dome digester

Chapter 3- Pic:4 Floating dome digester

The digesters in various biogas production systems may be operated either under Mesophilic (35-380C) or Thermophilic (50-550C) conditions. Mesophilic and thermophilic conditions involve different bacterial species. Digesters operating under mesophilic conditions are found to be safer and more stable.

Biogas production taking place during winters or in cold regions requires thermal insulation and/or heating of the digesters, as the optimal temperature for biogas production is between 35-380C. Lower temperatures lead to lower gas yields. At 150C biogas production is found to drop down to a very lower level.

Pre-Treatment

Biogas production is a process that is completed in several stages. The efficiency of the biogas production is seen to increase when a special chamber is also installed along with the digester to complete the initial stage.

If the raw materials used for biogas production are allowed to degrade partially before depositing in the digester, this will help in accelerating biogas production. This process is called Pre-treatment. Pre-digesters suitable for this are now available.

Chapter 3- Pic:5 Pre digesters

Pre-Digesters

BIOTECH India has developed various models of digesters suitable for pre-treatment. These are called pre-digesters and they are mainly divided into two categories, aerobic pre-digesters and anaerobic pre-digesters.

Aerobic pre-digesters are open tanks with specially designed filters. Anaerobic pre-digesters are designed in such a way that it is airtight. The manufacturing cost of anaerobic pre-digesters is comparatively higher when compared to that of aerobic pre-digesters as these are made from fiberglass.

Chapter 3- Pic:6 Anaerobic pre-digester

Chapter 3- Pic:7 Anaerobic pre-digesters connected with floating dome biogas plant

Chapter 3- Pic:8 Anaerobic pre-digesters connected with Non Return Valve system.

Chapter 4

Important Factors Affecting Biogas Production

Mother bacterial culture for initiating plant activity

Traditional model biogas plants using cattle dung as the raw material are found to take several days to become fully operational. To overcome this situation, biogas culture slurry can be used to accelerate the activity of microorganisms. Cow dung can also be used to activate the plant in situations where culture is not available.

In order for the biogas plants to start functioning smoothly, the microorganisms required at the initial stage of biogas production need to be grown inside the digester. In plants that do not use bacteria culture, biogas production may take a long time in the initial stages. Culture is used to avoid this. There are several methods that can be used to prepare the culture required for biogas plants. When biogas culture is required, the simplest way is to collect the slurry that comes out after treatment from a biogas plant that is functioning properly. This slurry is filled into the new digester that has to be operated. The slurry obtained from the biogas plant is rich in microbes that help in biogas production.

When many biogas plants have to be set up extensively in an area, the slurry from plants that have been running for a long time may not be sufficient. In such situations where a large quantity of biogas production culture is required, it would be useful to start a culture unit. A biogas culture production unit can be started in any area with limited investment. Biotech India has developed the technology for designing a biogas culture unit. Biotech also provides all the necessary technical know how and practical training for those who want to start their own biogas culture unit.

Prior to the development of biogas culture and special microorganisms, cattle dung was used to activate biogas plants. There are many basic factors to be considered for the smooth functioning of biogas plants. The efficiency of such biogas plants that are designed with a clear understanding of these factors will be very high. The biogas produced from plants in which all the factors are consistent will have a high concentration of methane.

The biogas plant will work efficiently only if factors such as temperature and pH inside the digester, structure of the materials used for processing,

Carbon-Nitrogen ratio, amount of waste treated daily, the amount of solid materials it contain, time period for which the bio-wastes remain in the plant etc are to be taken into account.

Temperature

Temperature inside the digester is one of the most important factors for better gas production. The microorganisms that operate inside the biogas plants require a certain level of temperature for the biological mixture in the digester to keep the microorganisms functioning. There are many different types of microorganisms in the digester, but constant heat is essential for the bacteria that produce methane. Slight variation in temperature inside the digester can also adversely affect their performance. Biogas production takes place mainly at two levels of temperature such as Mesophilic and Thermophilic temperatures . Biogas plants installed for domestic purposes and in public institutions operate at the mesophilic temperature.

Mesophilic temperature is about 35°C. Plants established in tropical countries do not require any special arrangements to maintain this mesophilic temperature. 25°C to 35°C is the optimum temperature for biogas production. When the temperature drops below 25°C, the biogas production from biogas plants is seen to decline gradually.

It is essential to check at regular intervals to ensure that adequate temperature is available for the smooth operation within active digesters. Industrial biogas plants are provided with monitoring systems for this purpose. Such machine-assisted monitoring is not practical in the case of domestic biogas plants.

The investment required for such monitoring equipments will probably be many times higher than the cost of domestic biogas plants. But through daily use and practical training, beneficiaries of domestic plants can quickly understand such things without the help of machines.

It is found that the process of the fermentation and gas generation process will be at the highest when the temperature is around 35°C. When the temperature drops below this level, the rate of fermentation is seen to reduce and the biogas plant produces only very little gas. Due to this, the gas production is seen to reduce considerably during winter season. The influence of temperature is less in the case of digesters constructed underground when compared to plants installed above the ground level. It may also be noted that the efficiency of the digester can be improved by raising the temperature inside the digester to a particular level . This level is known as the Thermophilic stage . To maintain Thermophilic level, additional heating has to be provided to the digester continuously.

Thermophilic temperature is optimal for processing more waste and producing more biogas in a short period of time on an industrial scale. Biogas plants in cold regions and biogas plants producing biogas on an industrial level operate at thermophilic temperatures. For this purpose, special heaters are installed in the digester to maintain the optimum temperature in the digester. Such plants will also be equipped with mechanisms for continuous turbulence of organic matter inside the plant to ensure even distribution of the heat to all areas. Thermophilic digesters require additional energy from outside.

Areas that are close to the equator are generally tropical. These areas have hot and dry climates for most of the year.

Biogas production in domestic biogas plants has been found to decrease during periods of extreme cold and when ambient temperature decreases due to continuous rainfall. This is because, as the temperature inside the digester decreases, the microorganisms become less active.

While installing plants in areas away from the equator and other relatively cold areas, it is best to provide insulation to prevent heat loss from the digester. In cold areas where plants are installed underground, as the surrounding soil acts as an insulator there is no heat loss from the digester.

A number of methods are used to control the temperature inside the digester. In power generation biogas plants, the heat generated during the operation of generators is passed through special pipes into the digester and circulated with the help of certain systems arranged inside the digester.

In situations where the digester needs to be heated from outside, the pipes required for this will be placed inside the digester during the construction of the digester. There will also be a facility to pass hot air through these pipes. Hot water can also be passed through these pipes instead of hot air when needed.

The deposition of hot waste and wastewater inside the digester also helps to raise the temperature inside the digester to some extent. However, in countries close to the equator, the climate would be ideal for biogas plants to operate at the mesophilic temperature. So there is no need for special insulation or any external heating. Therefore the construction cost of mesophilic digester will be comparatively low.

Amount of waste deposited in the plant (Loading rate)

The amount of waste deposited in the plant is an important factor for the smooth operation of the waste treatment biogas plant. Each plant has a maximum treatment capacity. Loading rate of a Biogas plant is the maximum treatment capacity of organic materials per day. It may vary depending upon the type of the organic material. The degradation time and the biogas production capacity of different organic materials will be different. When the loading rate is increased to more than the desired capacity of the digester, there is a chance for acid formation inside the digester and it will lead to the reduction in biogas production. Therefore while determining the amount of waste to be deposited in the plant; it should be according to the treatment capacity of the plant.

The slurry containing a certain amount of bio-waste and water will remain inside the digester where biogas production takes place. Only a limited number of microorganisms can grow in this area. There is a certain amount of waste that can be treated by those microorganisms. This amount is called the digester loading capacity. If the amount of waste deposited in the digester is lower than the loading capacity of the digester, there will be a proportional decrease in the biogas produced from the plant. But the plant will not have any other functional defects. However, if the waste deposited in the digester is more than the treatment capacity or loading rate, the activity of the plant is seen to slow down gradually.

Depending upon the size of each digester, their loading rates will also be different. Another factor to be considered while determining the loading rate is the structure of the organic matter that is to be deposited in the digester. The amount of biogas that can be obtained from each type of organic matter varies. For example, to produce 1 cubic meter of biogas, approximately 25 kilograms of cow dung is required. But one kilogram of fat can produce 1500 liters or 1.5 cubic meters of biogas. One kilogram of starch yields 800 liters or 0.8 cubic meter of biogas. In a 1 cubic meter plant, 25 kilograms of cow dung has to be deposited to obtain 1 cubic meter of biogas, however, if 25 kilograms of food waste containing starch is deposited instead of cow dung, one can only guess what the situation would be.

The loading rate of the plants should be determined by considering the size of the digester, the amount of waste deposited in it and their structure. The biogas generated from plants that are operated without considering such components is at the risk of being lost due to the inability to collect it or use it in a timely manner.

Time period for which the bio-wastes remain in the plant (HRT) Hydraulic Retention Time

The number of days or the time period for which the bio-wastes remain in the digester is also a factor influencing biogas production. The amount of biogas produced from the waste will vary depending on the period for which the waste remains inside the digester.

The retention period is the time for which fermentable material resides inside the digester. For generating maximum biogas from organic material, it should remain inside the digester for a particular period of time. It is observed that maximum gas production takes place within the first 3 to 4 weeks and then it decreases gradually.

Bio-waste can be fully treated only if it stays inside the digester for a certain period of time and produce the maximum amount of biogas available from it. When small digesters are installed or more waste than the treatment capacity of the digester is deposited in the plant, the waste entering the digester does not get the time to be fully treated.

From the moment a certain amount of waste reaches inside the digester, the microorganisms start working on them and begin producing biogas gradually till it reaches the maximum level. This time period will vary depending upon the structure of waste materials. The biogas production from bio-wastes is also found to gradually decrease after reaching its peak. However, these fluctuations in biogas production cannot be easily identified in biogas plants that are operated with a daily input of waste. The reason for this is that it becomes a cyclical process as new waste is being deposited into the plant every day.

In situations where slowly degrading bio-wastes are to be treated, the smooth functioning of the plant can easily be ensured by installing a pre-digester along with the digester.

Chapter 4- Pic:1 Predigester with digester

If the size of the digester is made smaller and the daily fed quantity of organic material is more than the capacity of the digester, then the treated slurry comes out as partially digested and is likely to attract flies and produce a bad odor. When this slurry is poured it is found to destroy plants. The same condition will persist even if waste more than the treatment capacity of the digester is deposited.

In the first few days, the waste entering the digester is found to produce more biogas and then the biogas production gradually decreases. The biogas production capacity of each type of waste will be different.

Some wastes decompose very quickly and produce more biogas quickly while others degrade only very slowly. Such wastes that degrade only very slowly needs to be retained in the digester for longer time periods. The attached annexure can be examined to understand more about the amount of biogas obtained from different types of waste.

Biogas from different sources of organic waste

Sl. No	Item	Quantity of waste	Biogas potential
1	Carbohydrate	1 kg	0.8 cum
2	Protein	1 kg	1.2 cum
3	Fat	1 kg	1.5 cum
4	Water Hyacinth	1Kg	260 Ltrs
5	Napier Grass	1Kg	90 Ltrs
6	Rubber sheet effluent	125 Ltrs	1 cum
1000 liters = One Cubic Meter			

Chapter 4- Pic:2 Table about the potential of biogas

By maintaining the temperature inside the digester, the retention time can be reduced considerably. The retention period of the feed material will also vary depending on the nutritional value of the organic matter. It is also noticed that the retention period could be considerably reduced if the temperature is raised to a particular level .This will also help to minimize the volume of the digester.

Role of pH

pH is a term which indicates the proportion of acidity and alkalinity of the feed materials. If the PH is lower, the acidity will be more and in high pH stage the alkalinity will be more. At a pH between 6.5 & 8.0 the gas formation will be the optimum. If the pH drops down below the desired level, the biogas production may get reduced or stopped. The excessive loading of the organic matter if continued, will lead to the growth of acid forming bacteria, resulting in lowering the pH. The percentage of methane in the biogas produced is found to be comparatively more, when the pH is between 7 – 8.

Fluctuations in pH have a significant effect on biogas production. This is because; changes in the pH cause variations in the chemical composition of the slurry which in turn affects the microbial balance and thereby causing fluctuations in biogas production.

Slight changes in pH within the digester can usually be resolved and brought back to equilibrium by the action of microorganisms without any specific action.

However, if acidity exceeds a certain level, the time taken for the plant to reach equilibrium will be more. By mixing a limited quantity of lime solution to the slurry and pouring it into the digester will help adjusting the acidity to some extent (5 gram for each kilogram of bio-waste).

Biogas production is also seen to be stagnant when alkali levels exceed a certain limit. To control this, limited amounts of hydrochloric acid can be mixed in the digester. This should be done only under the supervision of an experienced technician.

Biogas produced from low PH plants will have high carbon dioxide content and low methane content but in plants operating at high pH will have high methane content and low carbon dioxide content. The quality of biogas or fuel efficiency is determined on the basis of the amount of methane contained in the biogas.

Structure/ Composition of waste

The composition of the waste deposited in the digester is also an important factor for biogas production. Biogas production will vary according to the fluctuations in the basic elements contained in the waste.

The structure of the waste was not a major issue in those days when only cattle dung was used for biogas production. This is because, the organic matter is excreted as cattle dung only after it is partially processed in the digestive system of the cattle. The cattle dung which is excreted in this way will be rich in biogas producing bacteria. This is the reason why; the biogas plants using cattle dung as the raw material produce biogas without any other problems.

Cattle dung contains the right proportions of nutrients necessary to accelerate the action of microorganisms. This is why livestock dung is used as a basic component in the early stages of plant activation.

While installing biogas plants for domestic waste treatment, the waste can be deposited in the plant only after it is activated by using biogas slurry, culture or cattle dung mixture. Although the biogas production capacity of kitchen waste is much higher than that of livestock dung, an environment conducive for activation of the plant and the smooth functioning of the microorganisms must be initially created within the digester.

As household waste is a mixture of materials of different compositions and contains all the basic elements required for biogas production in the right proportions, the action of microbes is accelerated in domestic biogas plants also. Approximate amount of biogas available depending on the structure of the waste is given in .

Chapter 4- Pic:3 Domestic Biogas plant and customer

Carbon Nitrogen (C:N) ratio

Carbon nitrogen ratio is also a factor that plays a major role in biogas production. The ideal Carbon Nitrogen ratio for the biogas generation is 25:1. Carbon nitrogen content is found in this proportion in cattle dung. So cow dung is commonly used for the initial activation of the digesters in a cost effective way. The best range of C:N ratio is from 20:1 to 40:1. In this condition the microbes can grow so as to produce maximum biogas. Plant wastes are found to have high carbon content fish and meat residues and other animal and bird excreta have high nitrogen content.

As kitchen waste is a mixture of fish and meat waste and grain residue (carbohydrate) it is found to maintain this ratio to some extent. When carbon content is low, mixing dry plant wastes along with bio-waste and depositing into the digester will help in producing more biogas. Wastes containing carbohydrate also helps in maintaining this ratio.

Toxic substances that should not be deposited in the plant

Preventing the entry of substances that interfere with biogas production into the digester is just as important as considering various factors that accelerate bio-waste treatment. These mainly include toxic substances, pesticides, soap and soap products.

The presence of a few toxic materials in the feedstock will lead to reduction in the efficiency of the digester. Common toxic substances like fungicides, pesticides, insecticides, dettol, detergents, phenyl, heavy metals like zinc, copper, iron, nickel etc are harmful for the generation of biogas. If by chance, any of these materials happen to enter into the digester, the microbial organisms responsible for the gas generation will partially or fully be destroyed. If the biogas generation from a digester is stops due to the mixing of any one of the toxic materials, the digester has to be reactivated for further production of biogas. To avoid such a situation, one should not keep any toxic substances near the biogas plants.

Inorganic wastes/Non-biodegradable wastes should be completely prevented from entering the plant. This is not only because they cannot produce biogas but also because their constant accumulation can lead to further depletion in the efficiency of the digester. If chemicals and pesticides enter the digester it will lead to the destruction of microorganisms inside the digester. Biogas producing bacteria are those that operate under very controlled conditions. The extent of damage caused to them by chemicals depends on the amount and intensity of the chemicals that reach the digester. Biogas production will not be significantly affected if only very small amounts of soap or other chemicals get inside the digester. The plant will be fully operational in a few days even if biogas production gets reduced due to small amounts of chemicals. The duration will vary depending on the intensity of the chemicals which get mixed with the waste.

Digesters that have stopped producing biogas due to high levels of chemicals will remain inactive for a long period of time. At this point, new waste disposal to the digester needs to be stopped. If the damaged plant is to be quickly activated, the digester has to be re-activated using a new culture after removing the waste completely from the digester. For this, all the steps followed when a new digester is activated for the first time need to be completed. If due care is taken while collecting bio-waste for disposal into biogas plants, all situations arising when substances harmful to the operation enter the plant can be avoided.

Chapter 5

Biogas - A Clean Fuel

When compared to conventional fuels, it can be understood that biogas which is a source of renewable energy, is a clean fuel capable of replacing traditional / conventional fuels. Biogas has been categorised as a clean fuel as it does not cause any environmental pollution when used as a fuel.

It is to be noted that when compared to the production of conventional fuels, the production of biogas does not create any environmental problems. Another important feature of the biogas project is that the environmental pollution caused by the spread of biogas into the atmosphere can be mitigated to a certain extent by installing biogas plants and utilizing the biogas produced from it.

When conventional petroleum products such as petrol, diesel, kerosene, and fuels such as coal and wood are used for energy production, they are found to cause air pollution. The gases emitted when burning vehicle fuels are also harmful to health. Most of these emissions also accelerate climate change. The use of environment friendly fuels needs to be increased to control the increasing air pollution. Biogas tops the list of environmentally friendly fuels. Biogas production itself is a part of pollution control. If biogas had not been collected from the waste dumped in households and public places and from the wastewater, they would have decomposed and methane and other gases emitted from them would have spread into the atmosphere. Air pollution has been found to be significantly reduced when biogas is used as a fuel in diesel- powered engines.

Clean fuel from organic material

Due to the day to day activities, different types of organic wastes are being discharged from kitchens. These are cooked food waste, fruits and vegetables, fish and meat waste etc... These categories of waste are generally known as organic leftovers. The generation and discharge of organic leftovers on a daily basis is a major headache for the local self governments and the general public. Generation of clean fuel from organic leftovers is a good method for hygienic disposal of organic leftovers as well as for the management of these types of waste at the source itself. Bio-methanation method helps in managing organic waste in a decentralized way.

This zero pollution fuel will help to replace any other type of existing conventional fuels like firewood, cow dung cake, kerosene, petrol, diesel, LPG etc. Biogas can be used to recharge fuel cells also. Studies are still in progress to develop a cost-effective fuel cell, suitable for small and medium sized biogas plants.

Uses of BIOGAS

For cooking

Biogas generated from biogas digesters burn with a blue flame. This can be used as fuel for many purposes. Even though biogas has many uses, the main uses of biogas are cooking, lamp lighting, activating engines, electricity generation, and the manufacture of vehicle fuel.

The biogas produced through biomethanation is an efficient and effective alternative to conventional energy sources like firewood, kerosene, LPG etc. A Smoke free and clean kitchen can be ensured by using biogas instead of firewood / kerosene. Biogas helps to keep the vessels used for cooking, free from soot, because no carbon or smoke is emitted during the burning of biogas stoves. Biogas is safer and less dangerous, compared to LPG and other similar fuels used for cooking. The use of biogas is very economical as there is no recurring expense for the generation of it.

Chapter 5- Pic:1 Biogas Flame

Digesters of various models are used for biogas production. Digesters are mainly divided into three categories such as those used for domestic purpose, industrial purposes and commercial purposes. The amount of biogas produced varies according to the size of the digester, quantity and type of bio-waste deposited. Biogas generated from digesters used for domestic purpose is mainly used for cooking and lighting lamps. Suitable lamps, stoves and other ancillary equipments have been developed for this purpose. In Public Institutions like hospitals, schools, etc., biogas can effectively be utilized for boiling water and such other purposes like cooking .

Chapter 5- Pic:2 Biogas Light

Chapter 5- Pic:3 Domestic biogas stove (Single Burner)

Biogas stoves of various models suitable for domestic and industrial purposes are available. Biogas heaters suitable for heating water are also available. Various models of biogas lights which can be used for interior and exterior of houses are also available in the market. We just need to select the appropriate one based on our needs. The main use of biogas in domestic biogas plants is cooking.

Chapter 5- Pic:4 Institutional Stove

Every house having a biogas plant will be a powerhouse through the generation of bioenergy. Women in the houses can save a lot of time otherwise spent for procuring cooking fuel like firewood. People especially womenfolk can be safeguarded from various diseases like bronchitis, asthma etc and eye diseases caused due to regular exposure and breathing of smoke. The health and hygiene of such people can be upgraded by the efficient use of biogas.

Biogas stoves for cooking purposes are now readily available in various sizes and models. Of these, stoves having single burner, double burner and even having four burners are being manufactured and marketed by various companies. Biogas stoves are available in different models and at different prices, such as cast iron, stainless steel and glass top models.

Chapter 5- Pic:5 Different Model Stoves

Biogas brooders suitable for poultry farming and biogas driers for drying grains and rubber sheet using special biogas heaters are also now available.

Generation of Electricity from Biogas

The generation of electricity through conventional method is a costly affair. Severe environmental problems will be caused on account of electricity generation by utilizing coal, diesel etc. Even in the case of hydroelectric projects, environmental problems will arise in and around the project area on account of deforestation.

The generation of electricity using a biowaste treatment plant is not a costly concern. Of course the installation cost of a waste to electricity project may be a little high, but the recurring expenses are comparatively low.

Normally, 1- 1.25KW electricity can be produced from one cubic metre of biogas. Depending upon the percentage of methane content in biogas, the power generation may slightly vary. The size of the generator can be fixed based on factors such as the daily availability of biogas, time required for power generation and quantity of power to be generated. Power is generated by using biogas as the fuel in generators. Before feeding biogas as fuel into the generator the gas has to be passed through a gas scrubber to remove unwanted particles, gases, moisture etc.

Chapter 5- Pic:6 Waste to Electricity Project

There are two types of generators available for generating electricity from biogas. One is the dual fuel model which can be operated using two different fuels at the same time and the other is the 100% biogas fuel model which works fully on biogas.

Chapter 5- Pic:7 Waste to Electricity Project

Chapter 5- Pic:8 Dual Fuel Generator

Any fuel such as petrol or diesel is required for dual fuel model generators to start its operation. Once the generator starts its operation using any such fuel, biogas can be passed into the generator. Dual fuel generators normally work on an 80% - 20% mode such that it uses a mixture of 80% of biogas and 20% of the other fuel. When biogas is passed onto the dual fuel generator, the consumption of the other fuel automatically gets reduced to 20%. This is the reason why these generators are known as dual fuel generators.

For 100% biogas generators no other fuel is required either for starting or for further operation. This is the reason why these are known as 100% biogas engines. Dual fuel generators are mostly used only in biogas plants for domestic purpose and in biogas plants installed in institutions.

100 % biogas generators are mostly used in biogas plants suitable for public-private institutions and in big sized plants for industrial purpose. Institutions that develop such generators are operating worldwide.

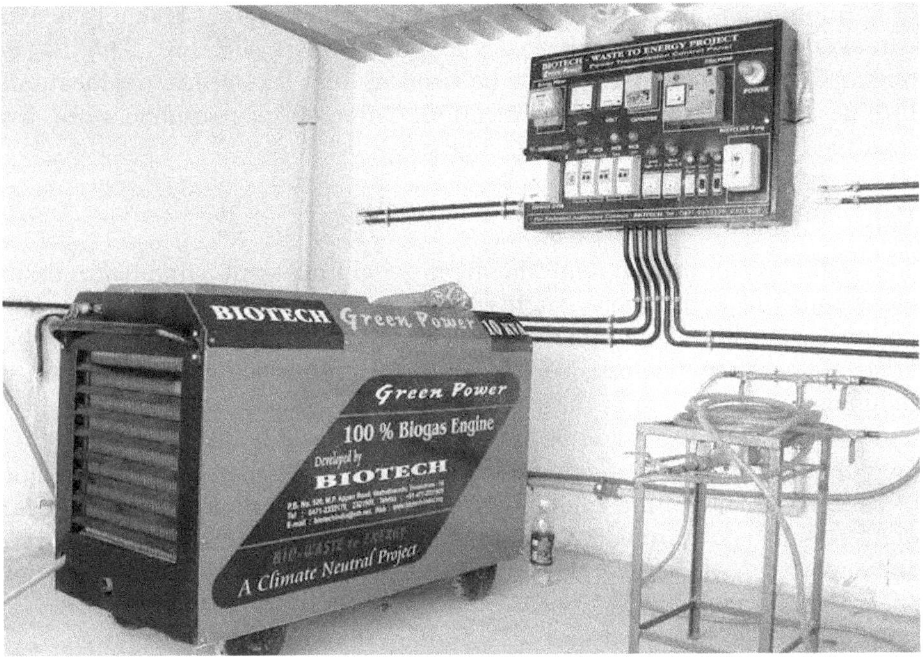

Chapter 5- Pic:9 100% Biogas Engine

All engines working on petrol can be converted into 100% biogas engines. Diesel engines can also be operated using 100% biogas by converting it into a spark plug model.

In areas where 100 % biogas generators are not available and for those who want to build a low cost biogas engine, vehicle engines can be considered. Petrol engines of vehicles that have been demolished due to malfunction or other reasons can be converted into 100% biogas engines with necessary modifications. This requires the services of an experienced engine mechanic. Similarly, the engines of old diesel vehicles after being redesigned into spark plug models can also be converted into 100% biogas engines. As the production cost is low, the beneficiaries will be satisfied even if their fuel efficiency is comparatively low.

Many companies have also developed micro turbines for biogas power generation. Various types of biogas turbines with high capacity ranging from 20 kilowatts are now available in the market. Although biogas turbines are the most expensive equipment in this range, their fuel efficiency is also very high.

For biogas power generation, stirling engines are also available. Stirling engines are mostly used to generate power from landfill gas at landfill sites. Stirling biogas engines are available in different capacities ranging from 20 kilowatts.

Another feature of biogas turbines and stirling engines is that they can run even on biogas containing very low amounts of methane. The biogas obtained from biogas plants can be used as fuel in sterling engines and turbines without any other purification. Therefore gas scrubbers are not required for the operation of such engines.

Vehicle Fuel

Vehicles running on natural conventional fuels emit a number of toxic gases into the atmosphere, such as carbon monoxide, which causes air pollution. Studies show that millions of vehicles emit toxic gases into the atmosphere every day, causing serious air pollution. Only by significantly reducing the emissions of these toxic gases can air pollution be controlled and the next generation will be able to live comfortably on this planet.

Tens of thousands of vehicles run on the streets every day. As a result of continuous usage, the availability of natural fuels such as petrol, diesel and kerosene is declining day by day. The search for the availability of non-conventional fuels gained strength with the decline in the availability of such natural fuels.

Natural gas was first used as an alternative to these. Natural gas has high fuel efficiency and no adverse effects during burning, like conventional fuels. For this reason, air pollution from vehicles using natural gas will also be less.

Biogas can be purified to the quality of natural gas. For this, all the other gases contained in biogas except methane have to be removed. Different technologies are used for the purification of biogas such as:

Chapter 5- Pic:10 Biogas Scrubber

Chapter 5- Pic:11 Water Scrubber

a. Water scrubbing method -Purification of biogas by passing through water.
b. Purification of biogas using special type of membrane.
c. Purification under high pressure through 'Pressure Swing Adsorption' technology known as P S A technology.
d. Purification by cooling biogas to very low temperatures which is known as Cryogenic technology.

These are the most commonly used technologies for the purification of biogas.

Purification of biogas takes place in different stages. The purification process involves the removal of carbon dioxide (CO_2), Hydrogen Sulfide (H_2S) and Moisture(H_2O) contained in the biogas.

Biogas produced in biogas plants can be upgraded to a quality that can be used as fuel for various purposes. If there is any residual biogas after being used for various purposes, the excess biogas thus produced can be treated and transported into the public gas grid. Many countries allow biogas to be purified and transported to the gas grid. If a special gas meter is installed for this purpose, it is possible to know how many cubic meters of biogas have been transmitted to the gas line.

Purified biogas can be used as fuel in all vehicles fitted with CNG - kits. Vehicles that are fitted with CNG kits do not require any type of modifications for this purpose.

When biogas is used as a vehicle fuel, the intensity of the noise produced as a result of operation of vehicle engines will also be less. Engines will function longer than they function on natural conventional fuels. Environmental pollution can also be significantly reduced as vehicle emissions are reduced.

In addition to using purified biogas as vehicle fuel, biogas can be utilized in many other ways.

Purified biogas is collected under high pressure in specially designed cylinders as is done in the case of CNG. The biogas collected under high pressure is known as Compressed Bio Gas (CBG). Compressed Bio Gas can be used as fuel for vehicles and as fuel for industrial purposes.

Chapter 5- Pic:12 Bio CNG Filling Station

Chapter 5- Pic:13 Compressed Biogas in Cylinders

In places where there are gas lines for the distribution of natural gas, purified biogas can also be distributed through these pipe lines for different uses. Carbon dioxide (CO_2) obtained during the purification of biogas can be used for agricultural purposes in green houses.

In biogas purification plants using cryogenic technology, Carbon dioxide is obtained in the form of ice. This is known as dry ice. This is can be used as an insulator for keeping medicines.

In gas scrubbers, hydrogen sulphide gets converted to sulphur. This can be collected and used for various purposes.

Chapter 5- Pic:14 Compressor designed for filling biogas in cylinders

Chapter 5- Pic:15 Compressed biogas storage in cylinders

Chapter 5- Pic:16 Biogas filling in vehicle

Chapter 5- Pic:17 Biogas purification system

Chapter 6

Important Parts of Biogas Plants

Important parts of portable domestic biogas plants are Digester, Gas collector, Inlet tank, Outlet pipe, Crusher, Biogas outlet, Gas control valve, stove, light, Additional weight and Scum breaker.

Floating Dome Model Portable biogas plant

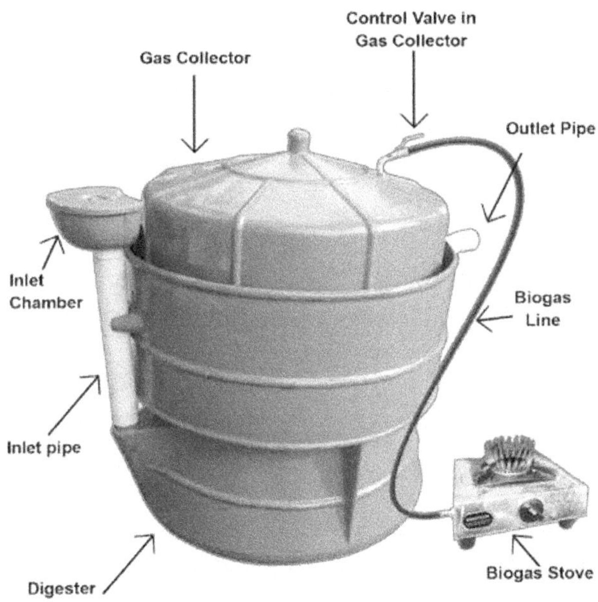

Chapter 6- Pic:1 Important parts of domestic portable biogas plant

Digester

This is a compartment or chamber that is not in contact with the atmosphere. This is the most important part of a biogas plant. Biogas is produced from the bio-waste deposited in the biogas plant through anaerobic fermentation process which takes place as a result of the action of microorganisms, in the digester. Different types of digesters are available depending on the model of biogas plants. Fixed dome model biogas plants are designed in such a way that the digester and gas collector are in a single chamber. In the case of biogas plants with moving gas collectors, the gas collectors are manufactured separately and later on attached to the digesters.

Bio-waste treatment and biogas production are equally important. The biogas production process gets complete only when the biogas produced from bio-wastes is collected without mixing with the atmosphere. Based on the plant model there are open digesters and completely closed (fixed dome) digesters. Open digesters are made from a variety of building materials, such as cement, steel and fiberglass. Once the construction of these digesters is complete, suitable gas collectors are attached to the open portion of these digesters. When these are attached, arrangements for the biogas produced in the digester to get collected in these gas collectors and to prevent it from leaking out will be made.

Fixed dome-Deenbandhu Model Biogas plant

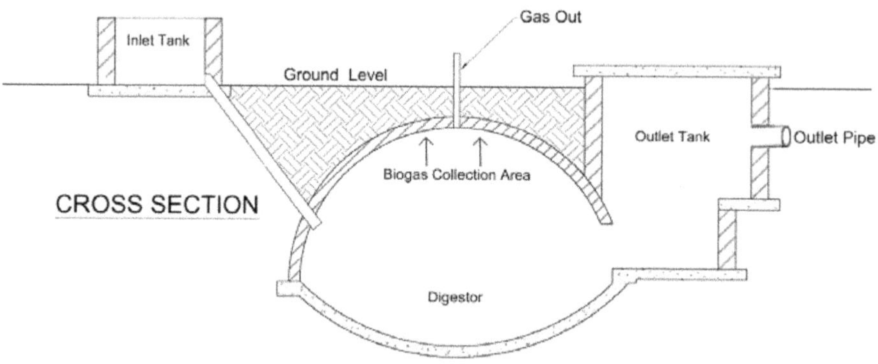

Chapter 6- Pic:2 Important Parts of Fixed Dome Model Biogas plant

Gas collector

The gas collector or gas storage tank is a provision for collecting the biogas generated in the digester. The gas collector may be an integral part of the digester as in the case of fixed dome model biogas plants or it may be a separate external arrangement based on the design of the plant. In the case of floating dome model plants, the gas collectors are mounted on the digester in such a way that it can move easily.

The gas collectors are constructed with a variety of building materials such as steel, iron, fiberglass and geo-membrane (synthetic fabric). Suitable gas collectors can be selected based on the model of the digester.

Chapter 6- Pic:3 Synthetic fabric (Geo membrane) gas collector

In cases where more biogas is to be collected, specially prepared additional biogas collectors made of building materials such as fiberglass and synthetic fabric are attached outside the digester. When these special collectors are made from fiberglass, a special tank is to be made with special frames inside it to which the gas collectors are attached.

Special gas collectors which are made of synthetic fabric also need a suitable protective covering.

Inlet Tank / Inlet chamber

This is a mechanism for feeding the organic waste into the plant. Inlet tank is an integral part of any model of biogas plant. In conventional model plants, this chamber is connected to the digester through a pipe. In advanced models, different types of mechanical feeding systems are available for feeding waste into the digesters of big sized biogas plants used for industrial purposes.

Inlet tanks of suitable sizes have to be constructed based on the model and size of biogas plants. In fixed dome plants, inlet tanks are constructed as a part of the digester using cement. There are inlet tanks that are specially designed for portable biogas plants. These inlet tanks are provided with a special cover and are designed to suit digesters of various sizes.

Necessary adjustments need to be made in the inlet according to the structure of the raw materials available for processing.

Crusher

If larger sized bio-wastes are available, a crusher also needs to be installed along with the inlet. The use of crusher to treat the waste and vegetable waste from vegetable markets and slaughter houses will help to increase the efficiency of the biogas plant.

Crushers are an integral part of industrial biogas plants.

Crushers suitable for domestic biogas plants are also available. Crushers that are suitable for treating kitchen wastes and that can be attached to kitchen sinks are also available. Crushers need electricity to function.

Crushers that do not require electricity and those that can be operated by hand are also available.

Crusher is not required in domestic biogas plants that treat only kitchen waste.

Outlet pipe

The treated material is discharged from the plant through the outlet pipe. In conventional model plants there is a chamber which is built in the vicinity of the digester to collect the treated material /bio-fertilizer discharged from the digester through the outlet pipe. In advanced model plants which are used for industrial purposes, there are different types of mechanical systems for taking out the treated materials from the digester.

The outlet for portable model biogas plants will be a pipe connected to the digester. There are also variations in attaching the outlet tank depending on the model of portable biogas plants. Outlet pipe will be attached to the top most portion of the digester in the case of ordinary portable biogas plants. There are also digesters made of fibreglass in which outlet pipes are constructed as a part of the digesters. In water jacket model portable plants, the outlet pipes are attached to the bottom portion of the digester.

Biogas outlet

Biogas outlet is the part that controls the transmission of biogas from the gas collector to the point of utilization. It is usually designed in such a way that it can be controlled by a valve.

Gas control valve

In the case of floating dome model biogas plants, the biogas outlet will be present at the top portion of the gas collector. In fixed dome model biogas plants, the gas outlet for supply of biogas will be fixed at the top most portion of the dome.

A gas control valve is required along with the gas outlet. A gas control valve needs to be attached to the gas outlet for collecting and storing the biogas produced in the digester in the gas collector itself. Keeping this valve closed can completely or partially control the flow of biogas from the digester to the gas line. Gas control valve placed on top of the digester is an essential element for the smooth completion of repair works related to the storage of biogas for digesters.

External weight

Minimum pressure is required to pass-on the biogas produced from the plant to the utilization point. A few models of biogas plants have the inbuilt facility to build up sufficient pressure in the digester itself. In such cases there is no need for additional weight. Other models need external weight for maintaining sufficient pressure. In floating dome model biogas plants, only if additional weights are mounted on the gas collectors, can the required amount of biogas be delivered to the intended use at the correct pressure.

Chapter 6- Pic: 4 Additional weight for BIOTECH MODEL gas collector

Biogas will be naturally stored at high pressure in fixed dome model biogas plants. For this reason, high pressure is obtained when biogas is taken out of the digester.

In floating dome plants and portable plants having gas collectors made of synthetic fabric, the biogas can flow out of the gas collector at the correct pressure only if a special pressure is applied to the gas collector. For this purpose special weight blocks of suitable sizes are kept on top of the gas collector.

As industrial-based biogas plants use special gas pumps for supplying biogas from synthetic fabric gas collectors, such large industrial plants do not require any special weight to be placed above the gas collector.

Pressure booster

In some biogas plants with geo-membrane gas collectors, external electrical pressure boosters and special pumps are used to pass-on the biogas with sufficient pressure. Pressure boosters are usually specially designed gas pumps. Pressure boosters help biogas plants to easily supply biogas to utilization points requiring biogas at high pressures. However, the fact that pressure boosters require electricity for their operation is a limitation. Pressure boosters that operate by using the electricity generated from solar panels are also available. Due to the shortage of electricity, pressure boosters that operate with the help of special storage batteries that store electricity are also available.

For operating machines such as biogas powered generators and compressors used for filling biogas cylinders, biogas needs to be provided at a certain pressure. Biogas coming out of ordinary model biogas plants will not have enough pressure to run high capacity generators. In such cases, specially designed pressure boosters are used to increase the pressure of biogas.

Chapter 6- Pic: 5 Pressure booster suitable for domestic biogas plants

Scum breaker

The processed and light weight organic matter that accumulates at the top surface of waste in the digester is called scum. The plants can begin to function more efficiently only if the scum thus formed is removed in a timely manner. The removal of the treated organic matter from the inside of the digester is essential to maintain the efficiency of a biogas plant. As special blades for breaking the scum are attached to the gas collector of floating model biogas plants, the possibility of scum formation in such floating model biogas plants is very rare.

In industrial biogas plants, as there is a continuous flow of waste materials, the waste materials in the digester are constantly kept in motion with the help of special circulation pumps. Due to this, the possibility of scum formation in such large sized plants is very rare.

Each biogas plant will have the capacity to treat a certain amount of bio-waste per day. Partially processed organic matter will accumulate on the surface of the digester when waste more than the treatment capacity of the digester is disposed constantly into the digester.

Small digesters that process fiber-rich organic waste are also more likely to accumulate organic waste after processing.

Waste accumulated after biogas production will lead to a significant reduction in the performance of the digester. Thus removing the scum that accumulates in the digester is a laborious task. But in floating dome model plants, the accumulated waste can be removed in a very short time.

In BIOTECH-made domestic biogas plants where scum breakers are available, it requires only a short period of time for the timely removal of bio-wastes that accumulate in the digester after the treatment process and for making the plant fully functional as before.

Arrangements for controlling the digester temperature

Although biogas digesters are available in a variety of models, one of the problems with their operation is that, the efficiency of the digesters subside with a decrease in the ambient temperature. Providing insulation around the digester is very effective in retaining the heat inside the digester. But as the settings required for this are very different, insulation is often found to cause inconvenience to the daily use of the biogas plant. As this is mandatory in industrial biogas plants, it is available as a part of the digester itself. In the tropics, digester insulation is not required, but during cold conditions when the ambient temperature drops below a certain level, necessary arrangements need to be made inside the digester to retain the heat.

Special heaters are available for biogas plants that are installed above the ground in cold regions for maintaining the temperature, providing special insulation and maintaining the heat of the bio-wastes in the digester.

Chapter 6- Pic: 6 Scum Breaker inside the gas collector of portable biogas plant

Chapter 6- Pic: 7 Heating arrangements inside of the digester.

Chapter 7

Design of Digesters

Design

The design of the digester is very important when preparing projects suitable for biogas production. The efficiency of biogas plants can be ensured only if they are designed to suit the structure of the available waste. Further servicing required by biogas plants that are properly designed, taking into account all the components, will be relatively less. Biogas plants should be designed only after considering various factors such as the kind of bio-waste that has to be treated each day, quantity of waste that has to be treated every day, whether the place where the plant is intended to be set up is a solid land or swampy area, the water table in that area, how low it has to placed to get water, the availability of water required for the plant operation, how much biogas needs to be collected in the plant.

There are different designs of biogas plants. Depending upon the requirements of the customers and the availability of the type of feed materials, appropriate designs can be selected. The common designs are single stage digesters, multistage digesters and digesters with pre-digesters.

Chapter 7- Pic:1 Scum inside the Digester

Scum, if formed in the digester can be removed. However, it is a laborious task and is quite expensive. Bio-wastes having high fiber content are more likely to form scum than other bio-wastes that are disposed in the plant. Digesters that can operate in multiple stages have been designed so that its performance is not affected by such waste materials.

Proper functioning of the plant can only be achieved by selecting a plant that is suitable for the available waste. For example, if bio wastes are deposited in a fixed dome model biogas plant which has been designed to treat cattle dung or human excreta, the plant is found to become inactive in a very short time. This is due to the fact that even if wastes having high fibre content contained in the organic waste are treated within the plants, the treated waste will accumulate on the surface of the digester and interfere with the smooth functioning of the plant. This is called scum. The scum if formed in the plant has to be removed. Removing scum from fixed dome plants is impractical and costly.

Single stage digestion

Different types of digesters are used to make biogas production more efficient. In single stage digestion, organic materials enter the digestion chamber of the biogas plant for fermentation process. After digestion, the organic matter in the plant will be discharged through the outlet of the same digester of the biogas plant. The whole fermentation process takes place inside the same digester. These types of digesters are being used commonly. Plants that are designed for domestic purposes also operate on the basis of single digestion. If such digesters are regularly overloaded, there is a chance for formation of acidity which may cause the plant to malfunction.

Chapter 7- Pic:2 Single stage digester

Although the biogas production process is completed in several stages, single stage digesters are designed in such a way that all of these stages can be completed within a single chamber. Single stage digesters are designed in such a way that, when bio-wastes are deposited in the digester, the newest of the wastes entering the digester goes to the bottom most portion and the oldest ones remain at the top surface of the digester. There will also be a partition wall inside the digester in the case of single stage biogas plants having water jacket. In this model biogas plants, both the inlet and outlet will be attached to the bottom of the digester.

The most important thing to consider while using single stage digesters is the amount of bio-wastes deposited in the digester. Single stage digesters will work efficiently for a long period of time if bio-wastes are deposited in accordance with the treatment capacity of the digesters.

Multistage digestion

In multistage digesters the organic materials will be passed through two or more digesters before being discharged from the plant. In some model plants, a few pre-digesters are also incorporated with the whole system. Multistage digestion takes place in plants having pre-digesters attached to it. The chance for the formation of acidity in multistage digester model biogas plants is very rare. The different stages of anaerobic digestion takes place in different chambers of these types of plants. The extraction of maximum biogas from the feed materials is also possible in multistage digester model biogas plants as the bio-wastes which reach the digester get treated completely. The quality of bio-fertilizer obtained from multi-stage digesters will also be very high. This can be directly used as fertilizer for plants.

Chapter 7- Pic:3 Multi stage Digester

The Multi-stage digester has been designed to easily dispose of all types of bio-waste that cannot be properly disposed of in a single stage digester. Multi-stage digester system can be carried out successfully even in small sized plants. Multi-stage digesters can be made in a variety of models. These can also be implemented by combining multiple digesters.

If excessive amounts of waste are deposited in conventional model digesters, it may lead to the presence of organic wastes that are not fully processed in the slurry that comes out of the digester after treatment. This leads to bad odor from the slurry. Such a possibility is very rare in multi-stage digesters. After processing in one digester the same organic waste is passed on to the next digester. At this point maximum biogas is obtained from the second digester. The same process takes place in the third digester also. Rich organic manure will be discharged from the third digester after all the remaining gas from the third digester is produced and all potential for biogas production is fully utilized.

Slurry de-watering machines are usually set up in projects where direct collection of solid organic manure is required. De-watering machines are expensive and require heavy power for their operation. For this reason, it is not economically viable to install such de-watering machines in small and medium-sized biogas plants. This is because this equipment can in many cases be more expensive than a biogas plant.

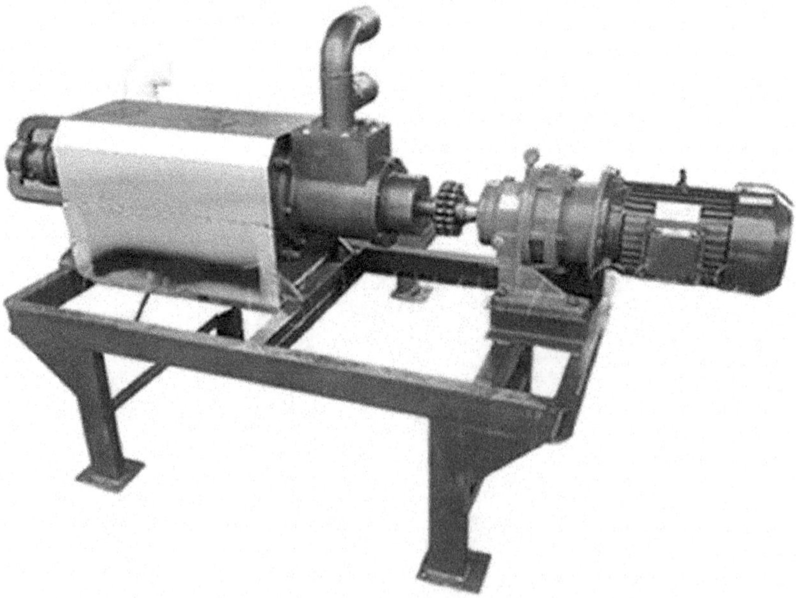

Chapter 7- Pic:4 Dewatering Machine

For plants up to 200 cubic meters, it is advisable to install pre-digesters instead of de-watering machines for the collection of solid organic manure. Only a short period of time is required for collecting processed organic manure from pre-digesters. The pre-digester which can treat up to 200 kilograms of bio-wastes daily, can be re-activated by removing the processed organic manure from it within a period of about one hour.

Chapter 7- Pic:5 Pre digester

Installing one or more pre-digesters along with the plant is sufficient if organic manure is to be collected at regular intervals after treatment from the biogas plant. Pre-digesters are not only a relatively inexpensive device to manufacture but also require no power to operate.

Classification of biogas plants

Biogas plants can also be classified into different categories based on their waste treatment method. All bio-wastes can be used for biogas production. Even though there are plants of different sizes available for non-industrial purposes, choosing the right type of digester for the raw materials used for biogas production will help to maximize the functionality of the plants.

Biogas plants can generally be classified into three main categories such as batch type, semi continuous type, and continuous type. This classification is mainly based on the type of feed materials used and their retention period.

Batch type digesters

High levels of biogas production are usually not possible from bio-wastes having high fiber content. Moreover, such bio-waste produces biogas very slowly. For a variety of reasons, depositing such organic waste into the digester on a daily basis is not practical. Batch type system is the one in which the organic waste materials are fed only in batches into the digester and no more feeding will be there till the end of the operation.

A digester can be reloaded with wastes only after the completion of biogas production and the treated bio-wastes are fully removed. These types of digesters are kept completely closed after being loaded with a batch of waste materials. In the initial stage it is seen that only very small quantities of biogas is produced and gradually the quantity of biogas obtained reaches the maximum level and then the biogas production is found to decrease.

On completion of the digestion, the digester will be opened to remove the treated materials to reload the digester again. In batch type, many digesters will be arranged one after the other in a single site to make sure the regular availability of biogas. The number and size of the digesters to be installed can be determined based on the amount of bio-waste received each day.

After depositing a batch of waste in one digester, it is closed and the bio-wastes that are generated afterwards will be deposited in the next digester. A minimum of 3 digesters have to be built in places where batch digesters are established to ensure continuous production of biogas.

Chapter 7- Pic:6 Bag type Digester

If at least three digesters are established, while waste is being deposited in one digester, the next digester will be at the peak of biogas production. At this point, biogas production in the third digester would have started to decline. The gas production would have stopped completely in the third digester by the time disposal of waste gets completed in the first digester and the digester is closed. At this time, the third digester is opened to remove the treated bio-wastes and again bio-wastes are to be deposited in this. This process will continue in all the digesters continuously.

This type of digesters would therefore require more investment and more space for its installation. Secondly, continuous availability of suitable quantity of feed materials also has to be ensured.

Often the crop residues left after the completion of the harvesting process degrade only very slowly. Their stalks will be very dry and strong. In many agricultural areas, there is a tendency to burn these to ashes and start cultivating the next crop. Studies in this field have shown that this can lead to serious environmental pollution. With batch type biogas plants, such hard bio-wastes can be treated and converted into rich organic manure without burning. Biogas produced during the treatment process can also be used as a fuel to meet various needs.

Semi continuous type

These are the models commonly used for the generation of biogas from cow dung and other organic materials. In this system, a predetermined quantity of feed materials mixed with water is charged into the digester from one side (inlet) at specified intervals, say once a day, according to the size of the digester/ treatment capacity of the digester. The chamber through which bio-wastes enter the plant is called the inlet chamber. The treated waste in the digester comes out in the form of slurry through the outlet line fixed at the other side of the digester.

The digested material (slurry) equivalent to the volume of the feed material will be coming out of the digester from the other side (outlet). In these models also the feeding of organic matter will take place at regular intervals, commonly two or three times a day. This type is best suited for households (family size), institutions (large sized) and villages (community size). The advantages of this type are, operationally these are more convenient with minimum labour, less cost and less space when compared to batch type, constant rate of gas production, less retention time, and more efficient decomposition. Formation of scum in the digester is the only disadvantage of this type. In family sized plants, one time feeding in a day will be sufficient for the smooth operation of the plant. Semi-continuous types are not very popular for field level operations.

Another specialty of semi-continuous digesters is that only one digester is sufficient to generate the required quantity of gas per day. The main components of a semi-continuous type biogas plant are, digestion chamber or digester, gas collector or gas holder for collection and delivery of gas, inlet arrangement for daily feeding and outlet arrangement for the discharge of the digested slurry, arrangement for gas distribution – pipe line, control valve for controlling the gas and appliances like stove, light, heater etc. for the utilization of the biogas produced.

There is a need to choose the most suitable plant in situations where bio-waste has to be deposited several times daily for biogas production. Although waste can be deposited in the plant several times a day, it is important to note that their quantity should not exceed the daily processing capacity of the plant.

Continuous type

These models are commonly used for the generation of biogas from liquid waste which is available on a continuous basis. This is mainly for the treatment of wastewater containing bio-wastes such as wastewater from slaughter houses, wastewater from food processing units and molasses from sugar factories. In this system, the feed material will be continuously discharged into the digester and an equivalent quantity of digested material will be continuously coming out from the digester.

New generation fully automated plants are also working on a continuous basis. There is no chance for scum formation in these types of plants. This may, however, need constant attention to monitor the temperature, loading rate, acidity-alkalinity ratio or pH etc and therefore large sized industrial biogas plants will be more suitable for continuous feeding of bio-wastes.

There is no doubt that waste water containing bio-waste is the most suitable raw material for biogas production. Wastewater flows out continuously from most of the factories which operate continuously for 24 hours. Food and grain processing units are good examples of this. Constant outflow of wastewater can lead to serious environmental pollution if not treated in a timely manner. Continuous type biogas plants are ideal for continuously receiving and treating waste.

Factors to be considered while designing a biogas plant

A biogas digester shall either be cylindrical, square or rectangular in shape. However cylindrical shape is preferred because it has a better load bearing capacity and it also helps in preventing pressure variations.

In cylindrical shaped digesters, the gas collector attached will also be cylindrical. This cylindrical gas collector will be designed in such a way that it can be moved up and down and also rotated to the sides. As scum breakers are attached to the gas collector, the scum in the digester gets broken down when the gas collector rotates.

The following points have to be considered while designing a biogas plant.

1. Volume of the digester depends on the type of feed material and its hydraulic retention time.
2. Volume of the gas collector depends on the expected gas output from the digester and the volume of gas required for storage and utilization.
3. Pressure of the gas delivered from the biogas plant to the utilization point .
4. Arrangement for heating, and insulation of the digesters installed above ground level in cold climatic regions.
5. Materials selected for the construction of plant should be strong to have a long life.
6. Methods adapted for the construction should ensure it leak proof for liquid as well as gas.

Domestic and industrial biogas plants have been designed in such a way that they can retain or store biogas produced for a maximum of 12 hours. Biogas collectors of suitable sizes are installed in the digester for this purpose. This means that the capacity of the biogas collector will be half of the daily biogas production capacity of the biogas plant. If extra biogas is to be stored for more days, special gas collectors suitable for this can be set up to store biogas as required.

Arrangements for the continuous availability of biogas produced by biogas plants that are installed for industrial purposes will be made in these plants. Therefore such large sized plants do not require special biogas collectors.

Proper study on feasibility before setting up a biogas plant will help to ensure smooth functioning of the plant in the future.

Practical knowledge in this field is of utmost importance. Biogas plants designed for bio-waste treatment should also be able to solve any future problems that may arise. For example, the biogas digesters that are installed for biogas production from food waste must be equipped with facilities to open the digesters for cleaning when required.

Unscientific design of biogas plants

Here, we have discussed about a few of the popular models of biogas plants commonly used. Many attempts are being made in different parts of the world to develop cost effective and advanced models of biogas plants. Many of these models are made by modifying different types of vessels into biogas plants. Almost all these models are working at a low efficiency due to different reasons including the leakage of biogas during the fermentation process. Methane, a major content in biogas is a good green fuel. The leakage of methane into the atmosphere will create serious environmental issues because methane is 22 times more dangerous than carbon dioxide. Therefore, designing of biogas plants should be done only under the direct supervision of technically qualified persons.

Chapter 7- Pic:7 Unscientific Digester

In biogas plants that are designed without adequate training, the bio-wastes are seen to accumulate after treatment. As a result of accumulation of waste inside the plant, the plant is found to become inactive within a few months of its initiation. In this case, the waste deposited at the inlet of the plant will be unable to flow into the plant. Resolving this problem requires a huge amount of money. To avoid such situations, the biogas plants should be designed in accordance with the structure of the waste.

The biogas plants that are marketed under the pretext such as completely concealed, odour free and compact will be a completely closed design. Therefore, such digesters cannot be opened up for examination. If such plants are established, it will result in a huge loss in the future, as the treated waste getting accumulated in the digester cannot be removed by any means.

Even multinational companies that offer innovative models of domestic biogas plants, market plants that cannot be opened up for servicing. Such plants will operate without any significant damages only for the first year after the installation of the plant. When these plants gradually become inoperable with continuous usage, the situation where they need to be discarded permanently may arise.

Chapter 7- Pic:8 Unscientific digesters

Chapter 7- Pic:9 Unscientific digesters 2

Those beneficiaries who select biogas plants for their own use should ensure that the plant they have selected is designed in such a way that it can be opened up for examination, whenever required. They should also consider whether the size of the plant is appropriate for treating the daily available wastes without producing any bad odour. Even if the plant is provided by renowned companies, the long term performance of the plants they provide must be ensured. The agencies that newly enter this field will have only one or two years of practical experience in this field. The waste materials start accumulating in the digester and start causing problems only two or three years after the plant is installed. These matters have been discussed based on the research conducted after establishing biogas plants of various models from the beginning of 2000. The biogas project, especially the domestic waste treatment biogas project will have endless business opportunities in the near future. Those who develop quality biogas plant models that have the ability to meet the needs of the people will have a bright future.

Chapter 8

Different Categories of Biogas Plants

As biogas production technology became more popular, it became available to all sections of the society. Biogas plants of appropriate sizes and models suitable for those who have bio-wastes are now available. The discovery that biogas plants are most suitable for decentralized waste treatment also led to rapid development in this field.

Different categories of biogas plants are available as per the requirement of the beneficiaries. Few of the important categories are domestic, Institutional, community and commercial ones. The suitable category of biogas plant can be selected based on the amount of waste available and the use of biogas and bio-fertilizer produced in the plant.

Domestic biogas plants

Chapter 8- Pic:1 Domestic Biogas plant

These categories of biogas plants are being used in domestic households. Under this category, plants of different capacities are available. Plants of sizes ranging from one cubic metre to six cubic meters are commonly used for domestic purpose. The size of a biogas plant is fixed based on the availability of feed material, the number of members in the family and the duration of the consumption of biogas. This does not mean that plants up to 6 cum can be used only for domestic purposes. Smaller establishments with fewer members may also need plants that are four to six cubic meters in size, depending on the amount of organic waste available there.

If domestic biogas plants are established with the aim of each biogas plant for each house, the tendency of throwing bio-wastes inadvertently from houses onto public streets and other open areas can be stopped permanently. All those who have their own bio-wastes can also become beneficiaries of this project. Majority of the beneficiaries are those who have taken the domestic connection for L.P.G. The use of L.P.G can also be significantly reduced by installing biogas plants in every household.

Institutional biogas plants

Huge quantities of organic waste are being generated in all Institutions due to day today activities. Based on the quantity of waste generated, biogas plants of appropriate capacities can be installed in all these institutions for the hygienic disposal of organic waste and for the generation of cooking gas.

Chapter 8- Pic:2 Institutional Biogas plant

Some of the potential Institutions are government institutions, semi-government and non-government institutions. For example, institutions such as hostels, factory canteens, hospitals, boarding schools, orphanages, old age homes etc. where many people reside and visit can treat their daily waste in a timely manner with the help of biogas plants. In addition, each institution can save at least 50% of their daily expenditure on cooking fuel.

As far as institutions are concerned, the cost of cooking gas has reached a level which is beyond what they can afford. The daily rise in the cost of cooking gas has become a huge financial burden for medium level institutions. If biogas is produced by utilizing all types of bio-wastes generated from these institutions, this can be used as cooking fuel. Wastewater containing a mixture of different types of bio-wastes and bio-materials is also obtained from each institution. By examining how much of this is available each day, the total amount of biogas that can be produced from this can also be understood. The investment required for setting up a waste treatment biogas plant in an institution will be got back within a maximum period of three years in the form of cooking gas and bio-fertilizer. Innovative schemes need to be devised and implemented so that such a useful biogas project can be turned into a popular project.

Community biogas plants

These categories of plants are commonly used for the disposal of organic waste generated from a group of families in a community. Community biogas plants are set up to treat the waste from different households together. Community biogas plants follow the method of disposing the waste received from each house and establishment into a plant which is placed at a convenient location suitable for all households in a village or a group. Bio-wastes are obtained in large quantities and biogas is also obtained accordingly.

Each household having organic kitchen leftovers or cow dung or any other organic waste available with them, can be collected together and brought to the plant for feeding. The biogas produced can be distributed to the houses through special pipelines, for cooking or lighting purposes.

If community kitchens are set up in conjunction with community biogas plants, people living in different households will be able to cook with the help of community kitchens.

The biogas produced through this process can be used to generate electricity. The electricity generated in this way can be delivered to each household and used for public purposes. Power generation through community biogas plants will always be a great boon to remote villages where electricity is not available.

Those living in remote villages may not have the financial means to install and use biogas plants on their own. The houses in such areas may be built in groups. Most of the villagers live in houses which are built in groups at certain distance away from open farm lands. Livestock breeding alongside rural households and agriculture will be the main occupation and source of income of the villagers. Biogas project can be used for uplifting the living standards of such people.

Those people living in remote villages and slums located near urban areas and those belonging to other weaker sections of the society will not have enough space to install biogas plants for waste treatment and cooking gas production on their own. But the timely treatment of their daily available wastes remains a major issue. Shortage of cooking fuel will also be a problem faced by them.

Community biogas plant is a project that can be considered for those people who face such a shortage in cooking fuel, who do not have access to scientific waste treatment facilities, who do not have enough space on their own to install waste treatment plants and who are financially backward. If the money required for the implementation of such projects is made available completely as subsidy or grant by the government, it will become a popular project that aims at the betterment of the society. The project can also be implemented efficiently by obtaining the money required for project implementation through sponsorship from private companies or from the social welfare fund (CSR) of corporate companies.

Commercial biogas plants

The realization that biogas can be used for many purposes in everyday life has given a huge impetus to the production of biogas. As a result, biogas production has grown into a profitable business. Larger sized biogas plants began to be set up for industrial production of biogas. It was the developed countries that gave more importance to industrial biogas projects. The launch of several new industrial biogas plants in other countries was also a good start.

Biogas plants can also play a significant role in the industrial growth of an area. Production of biogas on an industrial scale is a project that has high profit potential. Uninterrupted availability of raw materials has to be ensured for industrial biogas production. There are also companies in many developed countries that produce raw materials for industrial biogas plants. In industrial biogas plants, plant wastes are mostly used as raw materials for gas production. Due to the high biogas production capacity of such plants, they are cultivated on an industrial basis and is collected and used in biogas plants when they reach a certain growth. Cultivation of this type of plant is a good business in many countries. With the establishment of new industrial biogas plants, farms producing raw materials can be started as ancillary industries. Although all organic waste can be used as raw material in biogas plants, when biogas plants are set up on an industrial scale, suitable plant farms can be relied upon to ensure continuous availability of raw materials.

Chapter 8- Pic:3 Commercial Biogas plant

Commercial biogas plants mainly target the sale of biogas and organic manure. The biogas produced in a commercial biogas plant is commonly used for the generation of electricity. In addition, the biogas produced is purified to equal standards as that of natural gas and used as vehicle fuel. Purified biogas can be used as an alternative fuel in all vehicles running on natural gas.

The biogas produced from commercial biogas plants, after purification can be injected into the common gas grid. Natural gas supply grids have been established in various industrially developed countries. In such countries, the purified biogas from commercial biogas plants is found to be supplied to public distribution networks. The treated organic manure discharged from commercial biogas plants will be sold to the farmers in liquid or solid form.

Power generation by installing industrial biogas plants, transmission of purified biogas into the natural gas pipelines, vehicle fuel production are profitable businesses. In many countries, the producers of biogas are given permission to transmit the biogas produced in industrial biogas plants into the National Natural Gas pipelines. As an amount is fixed per kilogram, the manufacturers will get corresponding income from the biogas they produce.

If a biogas purification unit is established along with industrial biogas plants, biogas having quality equal to that of natural gas can be obtained. This can be used as vehicle fuel. In all the vehicles using natural gas as fuel, purified biogas can be used as an alternative fuel. Such industrial biogas projects that are established along national highways are one of the most environmentally friendly projects.

Factories producing high amounts of bio-waste can set up their own industrial biogas plants. This will also help them in reducing the use of other fuels.

Installing biogas plants will be profitable for big industries as the biogas is produced from their own raw materials and the biogas produced can also be used as fuel for their energy requirements. Sugarcane factories and flour mills are best examples for this.

Recovery of biogas from Sewage

With the growth of the society, the problems faced by the society also increased. Among these, one of the newest and one of the most discussed topics worldwide is climate change. Only some of the adverse effects caused by global warming and the resulting climate change are seen now. Many of the causes for global warming are the result of unscientific human lifestyle and unnecessary interferences with nature. Methane, one of the most important among green house gases and which plays a major role in global warming spreads into the atmosphere in different ways. It should be noted that methane gas is 22 times more dangerous than carbon dioxide. In view of this, it is essential for the very existence of mankind to take all possible measures to prevent the emission of methane into the atmosphere through man-made activities. Let us take a look at what the various man-made methane emissions are.

Methane emissions can occur as a result of cattle breeding and also from paddy fields, sewers and bio-wastes. Let us take a look at some of the things that can be done to control and reduce the rate of emission from these. It was in the mid-eighteenth century that the fuel efficiency of biogas was proven with the burning of gas from sewers. Since that period it has been discovered that biogas can be produced from sewage lines. Methane, which is a major component of biogas, is responsible for the burning of biogas. There will be several sewers or sewage canals or lines in each area. Millions of metric tons of biogas are released into the atmosphere every day from these sewage lines. This is because; the bio-wastes which reach the sewage lines or canals get accumulated at the bottom portion of the canals and biogas is mainly emitted from this. In many parts of the country, human waste is also seen to be discharged by directly connecting the toilet lines of houses into the sewage canals.

To control the amount of methane or biogas emitted from the sewers, the waste water before being discharged into the sewage lines can be passed through a sewage digester and the biogas produced in this can be collected. In this way, if wastewater is allowed to accumulate in digesters at various places and is discharged into the sewers only after collecting the biogas produced, the sewers that exist now will turn into streams or canals of pure water. Apart from this, another benefit is that the biogas which is produced from such waste water treatment digesters installed in various places can be purified and used as motor vehicle fuel or for the generation of electricity. Biogas when purified attains the same quality as that of Natural gas. Schemes for injecting this into the natural gas pipeline are being implemented in various developed countries.

Bio Septic Tanks

The idea of energy from sewage, capable of both bio-wastewater treatment and production of biogas which is a pure fuel should be made a reality. The developing countries are facing serious energy crisis. This project is most suitable for overcoming the energy crisis in such areas to a great extent. Apart from this, this project also indirectly helps in improving the health of the inhabitants in the area by reducing the rate of environmental pollution. Therefore there is an urgent need for a law to be enforced such that the sewage or wastewater released from all areas should be discharged into public water sources or canals only after it is treated in a biogas digester.

Chapter 8- Pic:4 Bio septic Tank

Like wastewater sources, septic tanks are also of importance in sewage treatment. Facts such as, how many millions of people are there around the world and the rate of increase in population each day can easily be understood by everyone. Majority of the world population consists of people who build septic tanks by digging a hole in the ground for human waste treatment or disposal of human waste.

In many places pre-fabricated or septic tanks prepared in advance are being installed. In areas where public drainage system is available, the human waste is discharged into drainage lines instead of collecting it in septic tanks.

Like any other bio-material, biogas is produced during the treatment of human waste also. It is understood that, from the excreta of about 30 to 35 people, 1000 litre or 1 cubic meter of biogas can be produced per day. Only when we understand how many people are there in the world can we realize the extent of biogas that can be obtained from human waste.

The wastewater reaching the sewage lines not only contains human waste. Water containing food waste discharged from the kitchens of households and establishments and also wastewater discharged from various industrial and food processing units reach the sewage lines directly. Many organisations, both big and small, that do not have their own wastewater treatment plants operate in various regions. They all depend on public canals for discharging wastewater.

Crushers that are capable of crushing the bio-waste along with water and discharging it in the form of a paste are being marketed in different institutions. In fact, these crushers are converting solid waste into liquid waste. The liquid waste thus produced is capable of producing high amounts of biogas. This also indirectly contributes to environmental pollution. Therefore a law should be enforced such that, if bio-waste treatment crushers are to be installed, biogas digesters should also be installed along with it.

Many areas have Public comfort stations or Public toilet complexes. Depending on the size and number of people using this facility daily, septic tanks of suitable capacities are also built along with this. The wastes that reach such septic tanks undergo biodegradation or fermentation and produce biogas. This mixes directly with the atmosphere. All the septic tanks in use in the toilet complexes will continue to release biogas into the atmosphere in proportion to the number of people using this facility. This is usually not noticed by anyone.

Biogas digesters can be installed along with all the public toilet complexes instead of septic tanks. The Biogas released from such digesters can be utilized for the generation of electricity. This gas can also be used for cooking purposes. Biogas produced from the toilet complexes can also be used for preparing the hot water required for cleaning the toilet complexes. Biogas can also be utilized as cooking gas by the households in that region. When all toilet complexes start using biogas digesters instead of septic tanks, the spread of biogas from such septic tanks into the atmosphere can be prevented completely.

Toilet Linked Biogas Plants

Toilet linked biogas plants are most suitable in places where there is not enough drainage facilities. Linking toilets with biogas plants is one of the best options for improving sanitation facilities in urban as well as rural areas. Toilet linked biogas plants also help in maintaining hygienic surroundings in slums and coastal areas. This is also a very practical and effective alternative technology.

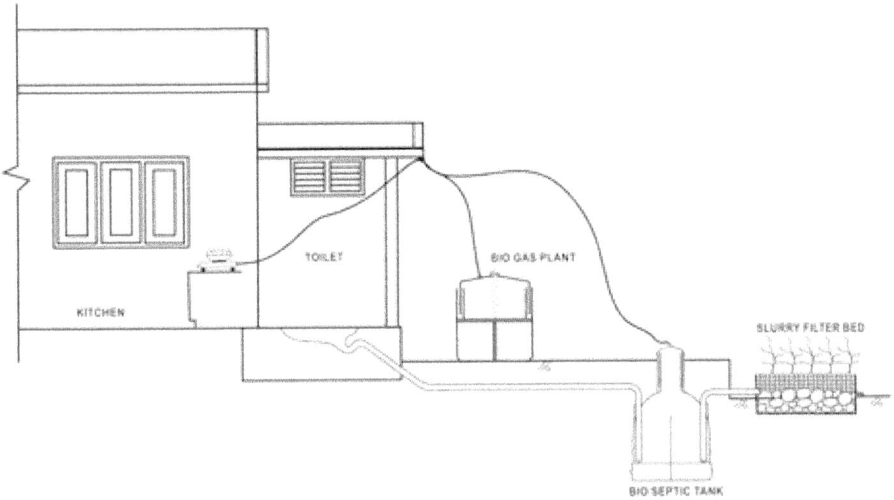

Chapter 8- Pic:5 Bio Septic Tank connected with Toilet & Biogas plant

Chapter 8- Pic:6 Bio septic tank designed for Institutions

Chapter 8- Pic:7 Toilet Linked Biogas plant

Toilet linked biogas plants use human waste as a source of organic material to generate biogas, which can be used for cooking, heating, and electricity generation. Toilet linked biogas plants offer several benefits, both environmental and social. Toilet linked biogas plants generate renewable energy from a source that is readily available in most communities: By using this waste as a fuel source, these plants reduce reliance on non-renewable energy sources such as fossil fuels.

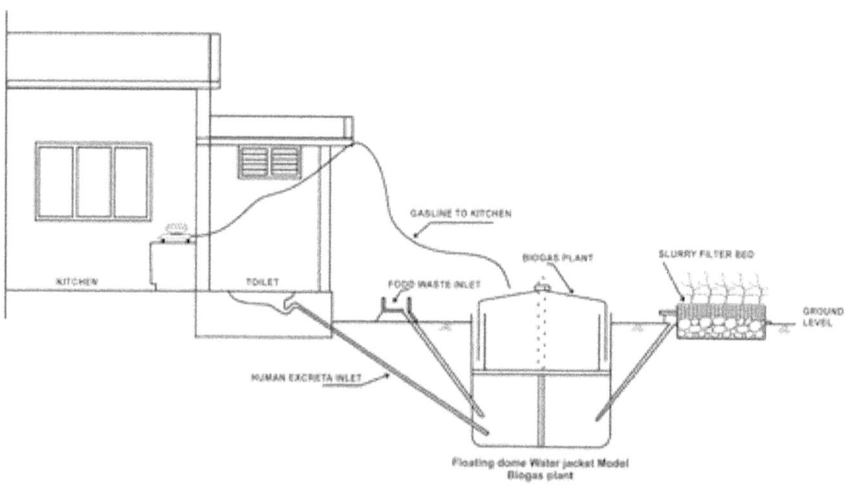

Chapter 8- Pic:8 Toilet linked biogas plant (Eco friendly Toilet)

Toilet linked biogas plants can help address the issue of untreated human waste in many communities. When waste is not properly treated, it can contaminate water sources and spread diseases. By using waste as a fuel source, these plants can help reduce the amount of untreated waste and prevent pollution.

Toilet linked biogas plants can be an affordable option for energy generation in many communities. They require relatively low initial investment and can provide a reliable source of energy for years to come. Building and operating toilet linked biogas plants can create new jobs in communities. These jobs can range from construction and installation to ongoing maintenance and operation. By reducing the amount of untreated human waste, toilet linked biogas plants can help improve the health of communities. This is particularly important in areas where access to clean water and sanitation is limited.

Chapter 9

Different Models of Biogas Plants

As a result of technological advances tremendous changes were happened in the design of biogas plants. The development of biogas technology from a combustible gas that can be extracted from sewers in the 18th century to a gas that can be produced in a Nano domestic biogas plant capable of becoming a part of any household is the result of centuries of research and hard work. The biogas project has grown so much as a result of the hard work of many researchers working in both the government and the private sector.

There are various types of biogas plants available in the market now to suit the needs of each beneficiary. With the development of the technology, many innovative models of biogas plants have started to become as a part of every home.

As a result of continuous research, various models of biogas plants developed over different periods have led to continuous development. The details of some of the most popular models of biogas plants have been documented here below.

Floating dome model

Floating dome model plants are of very simple designs having a cylindrical shaped digester and a cylindrical shaped gas collector fixed on the top portion of the digester. The gas collector is fixed on the digester with the help of a centre guide frame. This arrangement will give free movement to the gas collector. Once the biogas is stored in the gas collector it will move upwards and float on the top portion of the digester. When the biogas is used up, the gas collector will fall back down into the middle portion of the digester.

These models are known as floating dome model biogas plants because of the floating tendency of the gas collector. There are special categories of floating model biogas plants of various sizes installed for domestic and institutional purposes. They are 'water jacket model' and 'without water jacket model'. It can again be classified based on the utilization of raw materials for the construction of digester and gas collector.

The digesters are made mainly of Bricks, Reinforced Cement Concrete (RCC) and Ferro Cement Technology. The gas collectors of such plants are made of mild steel or Fibre glass Reinforced Plastic (FRP). In some models, a protective covering made from fiberglass is provided over the gas collectors made of iron to ensure long term performance. This helps to avoid painting the gas collector at intervals. Apart from this, the most widely used floating dome biogas plants are made entirely of fiberglass. The digesters and gas collectors of such plants will also be made of fiberglass.

Chapter 9- Pic:1 Floating dome model plant

In addition to these, fully pre fabricated portable floating dome model biogas plants made of Fibreglass Reinforced Plastic, HDPE, PVC etc. are also available. The most important feature is that it is very easy to fix any damages occurring to floating model biogas plants. Another feature of floating model biogas plants is that the biogas can be supplied to the utilization point at the same pressure from the beginning of biogas utilization to the time when biogas is completely exhausted.

BIOTECH Model biogas Plant

BIOTECH Model digesters were developed as a result of BIOTECH's ongoing research aimed at making biogas plants more popular. These are known as 'BIOTECH Models'. The rise in prices in all sectors has and had a major impact on the production of biogas plants. This plant was developed as a result of research into quality plants with a low cost of construction. Waste treatment is required in all categories of livestock centers. It is also an essential factor for farmers. A hygienic environment is highly essential for the efficient care of livestock. Farmers who rear livestock on a small and medium scale may not have much financial stability. Such people who find it difficult to even run their daily affairs properly may not be able to set up a waste treatment biogas plant at a high investment.

Chapter 9- Pic:2 BIOTECH Model plant

Cement tanks are usually built near the cattle shed for the farmers to collect cattle dung. An important feature of BIOTECH model plants is that such cement tanks can be modified and converted into digesters as required. Another feature of BIOTECH model plant is that, irrespective of the shape of such tanks, even if they are rectangular or circular in shape, can be converted into digesters. In biogas plants, the slurry usually flows out after treatment. In BIOTECH model plants, arrangements have been made for retaining the treated slurry in the digester itself for two or three days and for taking it out as and when required. Through this, the biogas storage capacity of the digesters can be increased or decreased according to the need of the beneficiary. Necessary arrangements for this can be done at the plant construction stage itself. Another feature of this system is that the digesters can be built in single membrane or double membrane models as required.

Highly skilled laborers are required for the construction of biogas digesters. But for the construction of BIOTECH model plants any person who knows the basic principles of construction work can easily carry out the construction of digesters. The required drawings are provided to the beneficiaries at the time of plant construction and the supervision by a technical expert is also available during the digester construction stages.

BIOTECH model plants are designed for those who require medium-sized digesters. Models of different sizes ranging from 10 cubic meters to 500 cubic meters are available. In special cases, BIOTECH plants of 1000 cubic meter capacity can also be constructed.

They have been designed in such a way that the biogas produced gets stored in the digester itself. An additional gas collector can also be built along with the plant to store the excess biogas produced so that it can be used by the beneficiaries as and when required.

If any type of servicing needs to be done in the digester, it can easily be carried out by detaching the gas collector from the digester. They have been designed in such a way that this can be carried out within a short period of time. The beneficiaries themselves can easily attach or detach the gas collector from the digester.

Excess pressure may be formed in the gas collector attached to the digester if biogas gets filled above a certain limit in the gas collector. The formation of excess pressure can adversely affect the long-term operation of the gas collector. A special safety device is also attached to the digester to drain out the excess biogas from the gas collector when excess pressure more than the permissible limit builds up in the gas collector. When excessive pressure is formed, the excess biogas so formed is seen to escape out in the form of bubbles through the safety device. Such a situation takes place only very rarely. This is because such a situation can be avoided if the biogas produced daily is used up the same day itself.

In BIOTECH model plants, gas collectors made of synthetic textile or synthetic fabric are used.

Another feature of BIOTECH model plants is that any damages occurring to the gas collectors can easily be rectified at minimal costs. The damages occurring to the gas collectors can be repaired as easily as the tubes of vehicle tyres. The raw materials required for this are now readily available in the local markets.

Two methods can be adopted for transporting the biogas which gets filled in the gas collector to the utilization point at the required pressure. Biogas gets stored in the geo-membrane gas collectors at a pressure close to atmospheric pressure. Necessary arrangements need to be made according to the plant model. In relatively small plants, additional pressure can be given to the gas collector by placing a special weight above the gas collector. As the pressure above the gas collector increases, so does the pressure of biogas coming out of the gas line. As a result, the pressure of biogas reaching the stove and other appurtenances will also increase.

The method of using special biogas pumps or pressure boosters for increasing the pressure of the biogas stored in synthetic gas collectors and then transporting it to the utilization point is also adopted. The only drawback of this method is that electricity is required for operating such pressure boosters.

The construction of BIOTECH model digesters can be carried out in three ways. They are; without water jacket model, with water jacket model and model with slurry collection tank. In without water jacket model, the gas collector will be attached inside the digester itself. The synthetic gas collector is attached to special hooks fitted inside the digester. In models with water jacket, a water jacket is also constructed at the top portion of the digester. The gas collector will be connected to this water jacket after it is filled with pure water.

In models without water jacket, the gas collector will remain immersed in the slurry inside the digester when the plants are in operation. As the biogas collector is mounted inside the water jacket in models with water jacket, it will be more convenient to complete servicing related works in situations where follow up servicing is to be carried out.

In digesters with slurry collection tanks, the slurry after treatment gets collected in special outlet tanks. These will have the facility to collect slurry for 2 or 3 days. Such model plants have been designed in such a way that the oldest slurry flows out from the tank when the slurry is not used up within this time span.

BIOTECH model biogas plants are most suitable for those who wish to install bigger sized biogas plants with minimal investment. BIOTECH model plants can be installed at a very low cost when compared to plants of other models having the same size.

Reinforced Cement Concrete (RCC) digester with Geo membrane gas collector model

These model plants are commonly used for bigger capacity plants for Institutional, industrial and commercial purposes. A unique feature of this model is that the digesters can be made rectangular or circular depending on the availability of space. The digester is constructed on site with Reinforced Cement Concrete (RCC) and the gas collector is made of special type Geo membrane (synthetic fabrics).

Chapter 9- Pic:3 Cement digester with Synthetic gas collector

This gas collector is fixed at the top portion of the digester with leak proof arrangements for the proper collection of biogas. Different types of RCC digester with Geo membrane gas collector model plants have been developed by various agencies.

Tube Model biogas plants

As the name implies, these are digesters made into a tube form with pre-fabricated geo-membrane or synthetic materials. These are constructed in such a way that air does not enter through either side of the tube. There will be a waste inlet pipeline at one end and a pipeline at the other end for the disposal of treated waste. In such models, the removal of scum, if formed will be a hectic task. For this reason, this model will be suitable for the treatment of wastes in liquid form or for the treatment of wastewater.

Chapter 9- Pic:4 Tube model Digester

They are comparatively cheaper models and can be installed easily, but the servicing and maintenance of these models after installation is not easy due to certain limitations in the design.

Tube digester is a model developed with the aim of minimizing the construction time and cost of a biogas plant. They have been designed in such a way that they can be easily installed even in areas where there is no vehicular access. Due to its light weight, the tube digesters can easily be transported through narrow passages to the installation site with the help of only one or two people. In such models, both digestion and gas storage takes place in the same chamber. Tube digesters are a model that can be used for simple experiments and in situations where the plant construction needs to be completed within a short period of time.

Lagoon biogas plants

Lagoon digester is an ideal model for factories that discharge large amounts of wastewater on a daily basis and for other establishments such as livestock and various animal farms. The use of biogas from lagoon digesters as fuel can also significantly reduce the use of other fuels. In the design of lagoon digesters, their size is not an issue. Any large surfaced sewage ponds can be converted into lagoon digesters.

Chapter 9- Pic:5 Lagoon Digester

For converting a wastewater pond into a lagoon digester, gas collectors of size suitable to the surface area of the pond can be made out of synthetic fabric. To build a more functional lagoon digester, a water jacket also needs to be constructed around them and the gas collectors should be attached to this. Lagoon digesters with water jacket are found to be more functional.

Large quantities of wastewater are regularly released from factories, slaughter houses and dairy farms working for agricultural growth. In most of the cases this wastewater contains huge quantities of organic material. This organic waste water is collected in big ponds for long periods.

As a result of the natural fermentation or treatment process, large quantities of biogas are seen to be released from these wastewater ponds. This gas will get mixed with the atmospheric air and pollute the environment considerably. But if a protective covering is provided over the wastewater collection ponds, the biogas emitted from these ponds can be collected and used as fuel.

The atmospheric pollution caused by the biogas emitted from these ponds can be controlled completely through this. Large sized geo-membrane gas collectors capable of completely covering such ponds are attached to the surface of the ponds such that there is no contact with the atmospheric air. This will help in completely preventing atmospheric pollution from such ponds and aid in environmental protection. As biogas is collected by providing a covering over the wastewater collection ponds, such models are called Lagoon biogas plants.

Janata Model biogas plant (Fixed Dome)

The Janata model biogas plants are very old models, which were commonly used in the early stages of biogas technology. The digester and gas collector of this model plants are made of bricks into a single chamber. The shape of the digester is partly cylindrical and partly spherical to minimize the cost of construction materials.

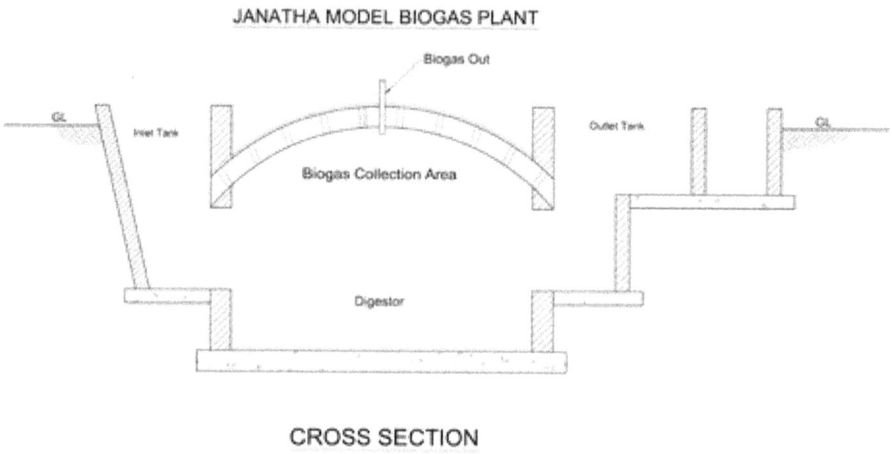

Chapter 9- Pic:6 Janata model

Due to the absence of a moving gas collector such models can be classified as fixed dome models. By analyzing the long term performance of such plants, these models were found not suitable for bio-waste treatment as there is no provision for breaking the scum formed inside the digester.

Even though these plants can be used effectively for the treatment of cattle dung, human excreta and wastewater, the cost of construction and the time needed for construction of such plants is comparatively high. When compared to other models, the installation of Janata model plants requires more space. Due to these reasons, Janata model biogas plants are not being used presently.

Deenabandhu model biogas plant (Fixed Dome)

Deenabandhu model plant is an improved version of Janata Model. Design modifications have been made to the digesters of such plants in such a way that the construction cost is reduced by minimizing the volume of the digester. The design of the inlet chamber has also been modified to minimize the construction cost and to prevent the loss of biogas through inlet chamber. When compared to portable model biogas plants, deenabandhu model biogas plants require more space for its construction, therefore it is difficult to install such plants in regions having limited space.

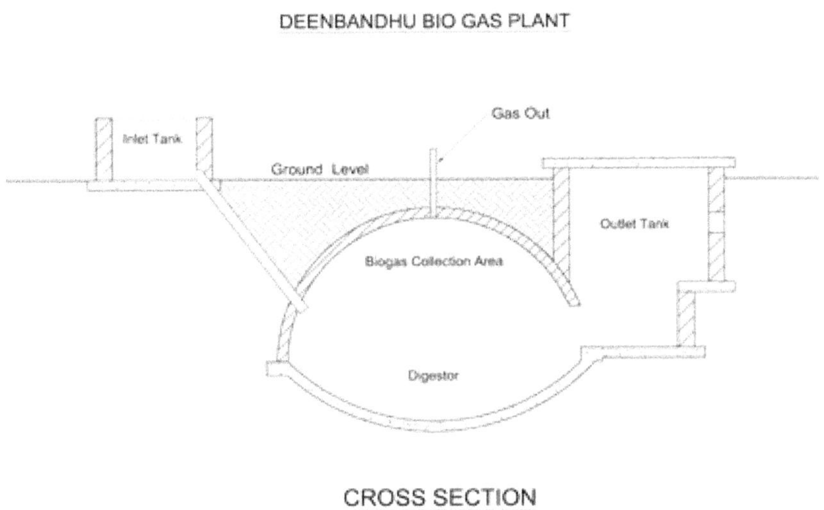

Chapter 9- Pic:7 Deenabandhu model

Most of the popular models of biogas plants come under the fixed dome category. One of the main features of such models is that the plant can be constructed using the raw materials available locally. The supervision of an experienced or highly trained technical expert is required for the construction of such plants. Deenabandhu and Janata model plants come under this category. A period of 7 to 20 days is required for their construction.

BIOTECH TURBO High pressure biogas plants

Biogas plants have already proven to be the most suitable equipment for domestic waste treatment and biogas production from all types of organic waste. With the change from cattle dung to kitchen waste as the raw material for biogas production mainly for domestic purposes, many people have become beneficiaries of domestic biogas plants and in many regions biogas plants have become an integral part of every home, like any other household appliance.

Chapter 9- Pic:8 BIOTECH TURBO Model Biogas plant

As the first one to develop portable biogas plants for domestic waste treatment, BIOTECH INDIA is constantly striving to improve the quality of such plants. With the increasing popularity of domestic biogas plants, BIOTECH research division is conducting a special study on their limitations and ways to overcome them.

One of the limitation is that the biogas obtained from the currently popular portable biogas plants is relatively low in pressure. Apart from this, it is also seen that the pressure of biogas decreases as the distance increases from the biogas plant to the place to be used through the pipeline. BIOTECH INDIA has designed High pressure biogas gas plants with 7 unique features to overcome both these limitations. Such plants are known under the trade name "BIOTECH TURBO High pressure biogas plants"

7 Unique features of BIOTECH TURBO High Pressure Biogas Plant

1. Fixed Dome Portable design – This unique design help to minimize the use of moving parts. It will provide durability and smooth functioning.

2. High Pressure Biogas – The availability of biogas with high pressure for hassle free cooking like LPG. It will help to minimize cooking time.
3. Automatic Slurry Mixing system – This will help to mitigate the formation of scum inside the digester. Through this feature more biogas can be produced within minimum period of time.
4. Automatic Pressure Regulator – This is one of the main advantages of TURBO biogas plant design. This will helps to prevent the formation of excess pressure inside the digester.
5. Detachable Gas Collector - For easy service and maintenance the digester and gas collector can be separated without much effort. It will help to complete the periodical service within minimum time as and when required.
6. Discharge Control System – The handling of treated slurry is one of the important task related to the regular operation of each organic waste treatment biogas plant. This unique feature helps to control the flow of treated slurry as and when required.
7. Centralized Feeding Provision - In conventional model portable biogas digesters the feed materials put inside the digester through the side bottom portion of the digester. This unique feature helps to keep input organic waste material inside the central bottom portion of the digester. It will help to accelerate the waste treatment process.

BIOTECH TURBO High pressure biogas plants are designed in such a way that any beneficiary can easily repair the damage. They are constructed in such a way that all parts of the plant can be stirred and changed within a very short time.

Biogas plants available in the market called pressure plants are designed to be closed and sealed so that they cannot be opened or repaired. If the waste accumulates inside such plants, there are no facilities to remove it. So the beneficiaries have to discard it after using it for a certain period of time. This results in heavy financial loss to the beneficiaries.

BIOTECH TURBO High pressure biogas plants are designed for longterm ease of use by the beneficiaries and the main feature of this plant is that the money spent on it is returned to the beneficiaries within a few years in the form of biogas and bio-fertilizer. BIOTECH TURBO High pressure biogas plants of different sizes can be designed to be installed on rooftops and open ground.

Floating Geo-membrane model (Ultra model)

This portable biogas plant in which the digester is made of fibre glass and the gas collector is made of Synthetic Geo-membrane. It has been designed to address the limitations of the existing pre fabricated portable models.

The plant has a number of features, such as more biogas storage capacity, limited space is enough to install the plant, the ability to control excess pressure and the incorporation of safety valves. This type of biogas plant is most suitable when considering biogas plants for food waste treatment. Another key feature is that there will be no open space in between the digester and gas collector. So there is no chance for mosquito growth in such model plants.

Chapter 9- Pic:9 Ultra Model biogas plant

Chapter 9- Pic:10 Ultra Model biogas plant with stove

Chapter 10

Categories of Biogas Plants

Different models of biogas plants have been developed by various agencies depending upon the geographical conditions and the availability of construction materials. Selecting plants suitable for each region will help in completing the construction or establishing the plants at a reduced cost. Portable biogas plants, pre-fabricated biogas plants and plants that are constructed at the site are the main models of biogas plants.

Portable biogas plants

Modern Technology makes many beneficial contributions to the public. Many innovations are made to find solutions to the problems faced by people. Redesigning a device that has been in use in the past should be done in a way that is convenient for customers to handle. But for the last one or two centuries, the situation has been such that the technology of producing bio fuels from bio-waste has gained popularity. Even though biogas projects became widely accepted by the middle of the 19th century, the equipments for the production of Biogas were not as easily available in the market as domestic appliances such as refrigerators and washing machines are now.

Farmers who used to raise their own cattle could use the Biogas produced from the cattle waste for their own purposes. Biogas was mainly used for purposes such as cooking, lighting lamp and for operating machines. Plants that produce biogas from cattle waste were later known as Gober gas plants. Mainly fixed dome and, floating dome model digesters were used as gober gas plants. It was at the beginning of the 20th century that the research into popular models of Biogas plants got accelerated as part of continuous development of technologies. The discovery that biogas could be produced not only from livestock waste but from all types of bio materials made the biogas project even more popular. It was at the beginning of the 20th century that such discoveries made as a result of many years of research and observations became more popular. Technology has developed to such an extent that anybody can produce Biogas if bio waste is available. The inability to dispose of bio-waste in a timely manner even during this period of industrial development has created many serious crises. Timely treatment of bio wastes generated on a daily basis from households and public-private institutions, without causing environmental pollution has become a matter of urgent importance.

Chapter 10- Pic:1 Portable Biogas plant

While taking advantage of the potential to produce bio fuel or biogas from bio-waste, those working in this field came to realize that bio-waste treatment is also becoming a reality. This increased the acceptance and demand for biogas production plants is tenfold. Local self governing bodies and local governments slowly started depending on biogas plants for processing bio waste in their respective areas. It was a great relief for the local bodies that the bio waste generated each day could be processed the same day itself, without piling it in one place. But such waste treatment biogas plants could be installed only by those who had adequate space for their installation. The increase in the population and the consequent increase in the number of buildings have led to the decrease in the availability of open spaces, day by day. The bio-waste generated from different localities used to be collected and brought together to a certain place and treated in the biogas plant installed there. The biogas thus obtained from the treatment of wastes was used for local purposes. This was mainly used for generating electricity for operating machines in the processing plant and for lighting lamps locally. The biogas obtained from these plants was also given as cooking fuel to the local inhabitants. The usage of the biogas was determined by the decision of each local body.

As the availability of daily bio-wastes started to increase, there was a situation where the wastes available could not be fully treated in the existing plants. In many cases, there was also a tendency for bio-waste to accumulate in public places. The technology of converting bio-residues into compost prevailed in most of the places before biogas plants were utilized for the treatment of bio-wastes. This is because of other modern technologies like incineration and plasma technology were not practical in financially backward areas. Due to serious environmental pollution in the vicinity of the areas where composting technology was implemented, local governments and private companies stated that using biogas production technology with relatively low environmental pollution for bio-waste treatment.

The idea of decentralized waste treatment gained importance when the existing centralized biogas plants became inadequate for bio-waste treatment. People also realised that with the availability of plants of different sizes suitable for decentralized bio-waste treatment, every household and institution could treat the bio-waste generated in their premises and use the biogas produced for their own use. Thus research into bio-waste treatment biogas plants suitable for both domestic and non-domestic purposes has been expedited. The institution, BIOTECH had already recognized the need of the hour for adequate biogas plants that were small in size and could be used like any other household appliance. The research wing of BIOTECH went ahead with the research in this field with the sole aim of developing a domestic biogas plant that is very simple, can easily be acquired and operated by a beneficiary on his own, light weight and durable. There were other important reasons for research into the simplest and most convenient biogas plants. The manufacturing time required for fixed dome plants and floating dome plants which were popular until the late 19th century, was about 7 to 15 days.

Such plants could be built only if well-trained and skilled construction workers were available. The availability of quality raw materials locally was also a major challenge. The plants built by untrained construction workers were often found to get damaged quickly or have a low performance from the beginning itself. Due to this reason, there were situations when the beneficiaries had to wait for skilled labourers. Though large numbers of workers were given training for the construction of Biogas plants at the government level each year, it was not sufficient to meet the present requirement. The large space required for the construction of such Biogas plants was also a problem. The ever-increasing number of dwellings also led to a decline in open spaces.

As the research into modern Biogas plants progressed, the limitations of existing plants were also a major issue. As result of the continuous research carried out by BIOTECH, a suitable biogas plant was developed in the early twentieth century to solve all of the above problems.

The existing Biogas plants were developed in such a way that it could be moved from one place to another when required. These plants later came to be known as portable biogas plants. When designing portable biogas plants, the main focus was on developing a model that could be easily operated. The first such plant was designed as a portable biogas plant based on the finding that floating dome models were most suitable for domestic use. The first portable biogas plant was unveiled at a press conference on August 18, 2000 to inform the public about the development of such a portable biogas plant. The first portable biogas plant had a digester of capacity 800 litres and gas collector of 600 litre capacity. As per the Intellectual Property Rights Act, BIOTECH was also able to obtain the design registration number, 221794/20/03/2009 for this plant.

Chapter 10- Pic:2 First Portable Floating Dome Model Plant

Chapter 10- Pic:3 Ordinary model

21 2000 ഓഗസ്റ്റ് 18 മലയാള മനോരമ

ഗാർഹിക മാലിന്യങ്ങൾ സംസ്കരിക്കാൻ പ്ലാന്റ്

Chapter 10- Pic:4 Press report about the invention of portable biogas plant

Floating model was chosen for the first portable biogas plant to make it easier for the beneficiaries to operate such plants. Fibre glass was used as the main raw material for the construction of such plants. The portable Biogas plants built with quality fibre glass will function efficiently for more than 20 years. The first portable biogas plant digester was designed by reducing the diameter of the plant and increasing the height so that it can be installed even by those who have minimal space. The base diameter of this digester was 80 cm. Different grades of raw materials are available for making products in fibreglass. High grade raw materials were used for the manufacture of portable biogas plants. Since biogas plants are in contact with water all the time, it is found that using low grade fibreglass can quickly disable such plants.

Bio-waste treatment capacity of the portable biogas plant that was developed first was 1000 to 1500 grams of food waste. A biogas plant of this size is suitable for the treatment of bio-waste in a nuclear family. These portable biogas plants are mostly suitable for a family of two to three members. Comparatively bigger digesters are needed for a family of more than three members. A 1500 litre capacity biogas plant is best suited for the treatment of bio-wastes from households of three to five members.

With the development of the first portable biogas plant, many people started installing such plants in their homes. With this, each of them got cooking gas from their own bio-wastes. Those who treated the bio-wastes from their own households with the help of domestic biogas plants received adequate amount of biogas to meet 50% of their cooking gas requirements per day. With the possibility of simultaneous waste treatment and cooking gas production, many were ready to install portable biogas plants in their homes. As a result within a very short period of time more than ten thousand beneficiaries started using domestic biogas plants.

With the invention that domestic biogas plant operation is possible with one's own bio-waste, people who generate large amounts of bio-wastes also started searching for portable biogas plants. As a result, portable biogas plants with the capacity to treat more Bio wastes were designed. After the 1500 litre or 1.5 cubic meter portable plant, the designers of BIOTECH designed portable biogas plants of different sizes such as 3 cum, 4.5 cum, 6 cum and 9 cum. The 3 cubic meter plant will have a digester of 2 cubic meters and a gas collector of 1 cubic meter capacity. Similarly, a 9 cum plant will have a digester of 6 cum and a gas collector of 3 cum capacity. The gas collector is designed in such a way that it can store up to 50 % of daily biogas production capacity of a portable biogas plant. That is, as the biogas produced during the day is fully utilized by evening, the biogas in the gas collector gets completely exhausted by evening. After this a biogas plant of this size is sufficient to store the biogas produced in the night till morning.

The next step in the design of portable biogas plants was to improve the quality of the existing portable plants. In floating dome portable plants, the organic matter after processing could be seen accumulating in the open spaces between the digester and gas collector. This was found to increase gradually and when excessive amounts were deposited around the sides, a situation would arise in which the smooth movement of the gas collector is hindered. This situation was seen in portable plants which treat food waste. But such a problem will not arise in portable plants using cattle dung as the raw material. However, the majority of the portable biogas plant beneficiaries use their plant for the treatment of food waste. For this reason, it has been decided to make necessary changes in the design of portable plants used for the treatment of food waste.

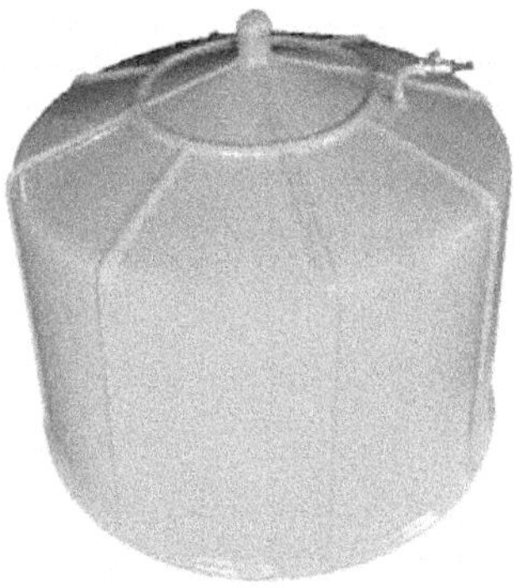

Chapter 10- Pic:5 Gas Collector of portable plant

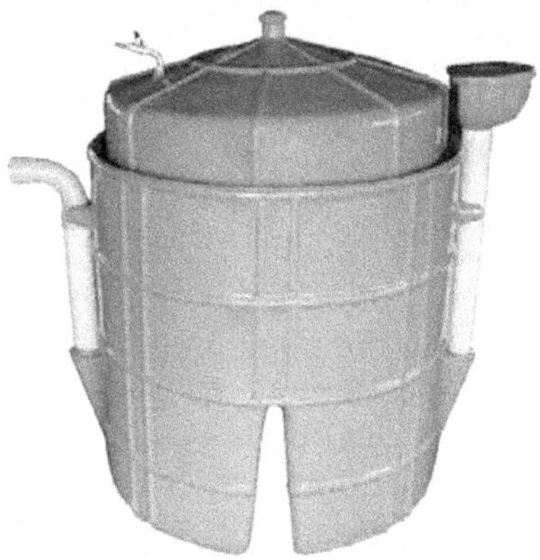

Chapter 10- Pic:6 Water Jacket model plant

Water Jacket Model Biogas Plants

Water jacket plants were designed by bringing about necessary changes in the design of portable floating dome plants. The most important feature of water jacket model is that the bio-materials after treatment are not visible through the gap between the digester and gas collector. For this purpose, a separate chamber was also built inside the digester from the topmost end to the middle section. The space below the middle section of the digester was divided into two chambers. The new separate chamber which was built on top is attached to the digester in such a way that no leakage takes place. Pure water or oil is filled in this chamber. The water jacket can also be filled with 90 percent water and any oil can be poured above this water. No expensive oils are required for this purpose. Instead, oils that are not of any use can also be used in water jackets.

Chapter 10- Pic:7 HDPE-Portable plant

In water jacket model floating dome portable biogas plants, the inlet pipe for depositing waste into the plant and outlet pipe for discharging the treated bio-wastes out of the plant are attached to the bottom portion of the digester. There will also be a wall built inside the digester to separate the two. This wall is known as partition wall. The partition wall is arranged in such a way that the waste reaching the digester is discharged out only after it is fully treated and after biogas is produced.

Portable biogas plants without water jacket are known as 'Ordinary models' and models with water jacket are known as 'Water jacket Models'. The cost of water jacket models will be higher when compared to the cost of ordinary models. Suitable models can be selected according to the need of the beneficiary and the biomaterials available for treatment.

There are several reasons behind the selection of fibreglass for the construction of portable biogas plants. With the widespread use of portable biogas plants, many plants will be needed every day. To meet this increased demand the availability of portable biogas plants in all areas needs to be ensured. The durability of fibreglass products is an important factor. High quality fibreglass biogas plants will work efficiently for more than 20 years. The raw materials required for the construction of fibreglass plants are available locally. There are a number of small scale industrial units that manufacture fibreglass products. If the moulds of portable biogas plants are made available to these units they can easily construct portable biogas plants. A very small investment is enough for an individual or an organization to start a portable biogas plant manufacturing unit. Portable biogas plants can be made of fibreglass, even in a small shed, with very little machinery and single phase electrical connection. This will also help in generating a number of employment opportunities locally. One of the notable features of fibreglass biogas plants is that any damages occurring to these plants can be fixed at a very low cost. It is also a fact that the construction of portable plants can be accelerated by the acquisition of fibreglass manufacturing units which are partially functioning or non-functioning due to inadequate employment opportunities.

When manufacturing portable plants in fibreglass, necessary precaution should be taken from the outset to maintain the quality of the products. While purchasing raw materials for plant construction, it is important to pay special attention to their grade or quality and select high quality raw materials. The use of marine grade or marine quality raw materials for the construction of portable plants will help in ensuring their long-term performance. When purchasing the raw materials such as resin, gelcode, glass mat, pigment, accelerator and catalyst used for the construction of fibreglass plants, the grading for quality should be examined and if materials of high grades are purchased, the plants constructed will also be of high quality.

The digester and gas collector of the biogas plants are mainly made of fibreglass. After ensuring the quality of the raw materials the next step should be to pay attention to the construction process. Depending on the grade and frequency of use of the glass mat used in the manufacture of fibreglass plants, the strength of the digester and gas collector may also vary. In different sized plants, glass mats of a particular grade according to the size of the plant should be used. For example, if a digester of one cubic meter or 1000 litre capacity is to be constructed, 450 grade glass mats should be used in two layers, but the product obtained by using two layers of 300 grade glass mats and that obtained from 450 grade mats will look the same. But there will be a big difference between them in terms of strength. Plants that are built without enough glass mats are more likely to have leaks and cracks in a short period of time. It is also essential to pay special attention to these matters during construction and give the manufacturer the necessary instructions.

The time is not far off when portable biogas plants will become an integral part of every household. With the expansion of portable biogas plants, these plants have become available in many places in a way that allows beneficiaries to choose the right plants for their needs, just like any other home appliance. The beneficiaries can buy and install these plants which are very lightweight and compact to meet their household needs in a convenient manner. While selecting a suitable plant, the most important thing to keep in mind is their quality. The best solution for this is to buy portable plants only from manufacturers who have proven their operational excellence. It is also important to keep in mind that investing in substandard plants considering only the initial investment and profit; it will only result in a loss of investment. As the project progresses with the goal of having one portable biogas plant in every household, the necessary practical training and guidance needed by those who want to start manufacturing portable biogas plants will be given by BIOTECH.

The design of portable biogas plants have led to a major breakthrough in the field of domestic biogas production and increase in the number of beneficiaries using biogas plants. Portable biogas plants have been designed for handling plants conveniently and for shifting them from one place to the other, if necessary, without much effort.

Portable biogas plants have been designed in such a way that it can be used effortlessly like any other household equipment.

Portable plants are mainly produced using Fibreglass Reinforced Plastic [FRP], HDPE, PVC etc. The quality of portable biogas plants can be maintained only if the plants are designed in accordance with the type of raw material used for plant construction.

Chapter 10- Pic:8 Pre fabricated Cement digester with fibreglass gas collector

To ensure the quality and long term performance of portable biogas plants, the plants have to be designed based on the type of raw material used.

Prefabricated biogas plants

In these models of biogas plants, the whole plant or different parts of the plant are produced in the factory and kept ready for delivery. The final assembling of these plants can be completed at the installation site in a very short period of time. One of the main advantages of this type of plant is that the construction of the plant can be completed very quickly.

It is more profitable to complete the construction of large biogas plants with high efficiency in suitable industries itself. Such fully constructed plants can be disassembled into different parts and then reassembled at the construction site which can significantly reduce the construction time. Depending on the size of the plants, they can be moved from the factory to the construction site as a single unit or after dividing it into different parts.

Pre-fabricated biogas plants are relatively expensive to manufacture but as far as the beneficiaries are concerned the plant construction can be completed in a very short period of time. Prefabricated plants can be relocated from one place to another as and when required.

Biogas plants Constructed at the site

There are also plants that are constructed by procuring and using the raw materials for biogas plant construction from the construction site premises. Although the model of such plants is the same, the raw materials used for their construction will be different. Fixed dome model plants such as Janata model and Dheenabandhu model and digesters of industry- based large sized plants are constructed using such locally available construction materials. Janata and Dheenabandhu model plants are constructed with bricks, sand, cement and rocks. But in areas where brick is not available, cement blocks are used instead of bricks for the construction of the digester. Ferrocement technology is used for their construction in some other areas. Larger sized fixed dome plants are also constructed using Reinforced Cement Concrete (RCC).

Chapter 10- Pic:9 Fixed dome Model

Chapter 10- Pic:10 Biogas Plant construction with central block

Chapter 10- Pic:11 Construction of Floating dome model biogas plant

The digesters of floating dome model plants can also be constructed in the same model using locally available construction materials. Utilizing locally available construction materials is more convenient in situations where many plants need to be set up in a short period of time in distant rural areas and areas with low vehicle transport facility. The development of such plants can be accelerated by selecting and giving adequate training to those who are interested in job training locally. This will also help in creating more jobs locally.

Some of the popular models of biogas plants are constructed at the site itself. These models cannot be fabricated in advance. One of the main advantages of this model plant is that it can be constructed with locally available materials. Specially trained persons are highly essential for the installation, assembling etc of this model plants.

Chapter 10- Pic:12 Anaerobically treated solid waste

Application of Biogas Plants

Biogas Plants can be used for different applications, such as hygienic disposal of organic waste, generation of green energy, production of organic fertilizer etc. Biogas plants can be installed in rural villages to provide clean fuel by, replacing the use of conventional fuels like firewood, cow dung cake etc. It can also be used to improve the industrial sector by using the biogas produced as a fuel for industrial applications. Biogas plants can be used to generate electricity in remote villages where there is no grid connection. With the installation of domestic biogas plants in rural villages, a lot of job opportunities can also be created for the benefit of the villagers.

Sanitation facilities can be improved in slum areas through the installation of toilet linked biogas plants. Organic waste disposal issues in slum areas can also be overcome through the installation of decentralized biogas plants. The coastal areas are thickly populated and most of the houses in these areas are constructed without proper sanitation facilities. To overcome this situation installation of family size biogas plants connected with toilets are the best option.

Chapter 10- Pic:13 Anaerobically treated liquid waste

Chapter 10- Pic:14 Waste to electricity project

Chapter 11
Nano Biogas Plant

In a situation where energy production from biomaterials has become inevitable, there is a lot of research and experiments going on around the world in this field at various levels. People conducting such research include scientists, research students as well as the general public with an interest in science. Most of this research was conducted in laboratories where research facilities were available. But only a small section of the population is able to do this kind of research. Due to lack of adequate testing facilities, many people who are interested and talented in such research are unable to put it into practice. Providing the infrastructure for research enthusiasts to work creatively at the level they intend will enable them to make many innovations at a faster rate and at higher standards than they currently are.

As far as the common people are concerned, the production of Biogas from bio-waste is an innovative idea. In order to give people proper confidence in this technological advancement, display of such working models / demonstrations and training programs need to be organized. Functional biogas plants will always be more beneficial for such training programs. The establishment of such functional exhibition plants in each area for the benefit of the people is a matter of huge financial liability. In the present scenario, it would take a lot of time to set up such exhibition plants in each place, make it functional and then use it for the people of that area. This is something that hinders the time-bound expansion of such innovative technology. It would be very helpful for those involved in the expansion and promotion of the biogas project if exhibition plants are available in sizes and models that can easily be transported from one exhibition site to the next.

For those working in the field of biogas and common people who want to do research in this field, there is a widespread tendency to buy locally available barrels and transform it into digesters when suitable biogas digesters are not available for testing. Not only is the efficiency of these digesters low, but the risk of leakage of gas produced from such digesters is also very high. These digesters will not have the facility for the easy disposal of bio-materials into the digester and for the removal of the processed material out of the digester. This often becomes an obstacle for the smooth progress of research. The gas that leaks out from such digesters causes serious environmental pollution.

The biogas storage capacity of such digesters installed by each individual and institution, based on their convenience, will also be comparatively low. Due to these reasons, the experiments cannot be completed on time and to the expected standards. Since most of these digesters were fixed dome models, it is difficult to determine the amount of biogas stored in such digesters. A situation often arises where accessories and component parts (spare parts) suitable for the model of experimental biogas digesters and of appropriate sizes are unavailable. For this reason, researchers are compelled to build digesters using materials of sizes available in the market. This leads to a decrease in the efficiency of the experiments they conduct.

The research division of BIOTECH initiated efforts to design biogas digesters that are suitable for biogas project research and for conducting awareness programs under the motivation and demands from those working in this field and organizations that promote green energy projects. The main focus while designing this type of digester was to create a model that was simple and easy for anyone to use. From preliminary studies, it was found that while designing digesters for experiments, digesters that are suitable for laboratories and those suitable for display were required. The long-term reusability of such digesters was also considered while designing digesters that are suitable for both these needs.

Chapter 11- Pic:1 Nano 45 Ltr Digester

A lab-sized biogas digester was the first to be considered in the design of digesters suitable for practical training. For this, it was decided to design a digester with a storage capacity of 30 liters and a gas collector with a biogas storage capacity of 15 liters. In the design of research digesters, special consideration was given to the model of the digesters.

Since the operation of floating dome model biogas plants was highly efficient, it was decided to develop digesters of this model which is suitable for conducting experiments. Such experimental digesters were constructed in the floating dome model, since the assembling and dismantling of its component parts could be completed within a very short period of time. An important feature of the floating dome model is that any damages to these digesters can easily be rectified within a short period of time. This will help in completing the experiments on time without any interruption.

Another important factor considered in the design of experimental digesters was the selection of raw materials for their construction. Exhibition digesters often need to be shifted from one place to another.

In such situations where the digester needs to be constantly moved from one place to another, there is a high possibility for the digesters to get damaged. Therefore strong and durable raw materials have been used for the construction of these digesters such that they can function efficiently over a long period of time without any damages occurring rapidly. The raw materials that are available locally were selected for repair so that the beneficiaries themselves could repair any damage in a very short period of time. Taking into consideration these factors, it was found that fiber-glass reinforced plastic or FRP would be suitable for the construction of experimental digesters. These are known as 'Fiberglass digesters' in short.

Another important feature of these digesters made of fiberglass is that they are comparatively lighter in weight and therefore easier to handle. When a digester model was selected for designing experimental digesters, fiberglass was found to be most suitable for the construction of digesters.

Digesters for Experiments

The name "Nano digester" was given to these digesters considering their size. As the name suggests these digesters are of micro level and can be handled very easily. Nano digesters are known as "Nano Biogas plants". The prototype developed primarily for the design of Nano Biogas plants is available under the name 'Nano 45'. As the name suggests, their total volume is 45 liter. Why it was not 50 liters is also a relevant question. The design wing of BIOTECH gave primary consideration on how to control the problems that may arise during the transportation of these. The packing size of Nano 45 model plant is 2 cubic feet. In more detail, as space is an important factor while transporting the packaged plant to other places, the Nano 45 model is designed to fit in a 2 cubic feet sized packing. Nano 45 can be packed in a package having a height of 60 cm and length and breadth of 30 cm.

The outer covering of the digester and gas collector is made from fiberglass and the metal parts inside are made of galvanized iron. The floating dome model biogas plants have been designed in such a way that their working principle can easily be understood. With each Nano 45 model digester, a 150 cm gas line, gas control valve outside the gas collector, hose collar to be attached to the valve, hose clips for connecting the gas lines, experimental lamps for testing the fuel efficiency of Biogas by burning, additional weight to be placed above the gas collector, inlet tank of the digester, cover for inlet tank, cover for the additional weight placed on top of the gas collector are available as basic equipment.

BIOTECH also provides a special biogas culture developed by its research division, which is rich in microorganisms and is to be used during the initiation / activation of the digesters along with these digesters. Moreover, this culture also helps in accelerating the production of biogas. The time required for the production of biogas in the initial stages can be reduced considerably by using this culture. The treated slurry coming out from biogas plants operating locally without this culture can also be collected and used for initiating Nano biogas plants. In situations where slurry from biogas plants is not available, digesters can also be operated by using cattle dung. But when cow dung is used, it may take more time to start biogas production in the early stages. The duration will vary depending on the quality of cow dung used.

Another important factor under consideration at the time of designing a Nano digester was how to handle a situation where the plant had to be moved from one place to the other. A fully functional digester could be moved from one place to another only after the bio materials inside the digester were completely removed and cleaned properly. To overcome this limitation, a platform with special wheels was made to place the digester. The digester can easily be moved from one place to the other if it is activated after placing it on this platform.

When the platform was designed as a trolley model platform with handles, it became easier to shift a fully operational plant from one place to another. Such trolleys were found to be more convenient in situations where a functional digester was to be transported in a vehicle.

The maximum amount of biogas stored at a time in a Nano 45 model digester is 10-15 litres. It would not be practical to operate a stove burner with this much quantity of biogas. As a solution to this, an experimental burner used in the laboratories was included along with the Nano 45 digester. As experimental burners require only small amounts of biogas, experiments can be performed for a longer period.

Nano 45 model Biogas plants are mainly used for research purposes. They have been designed in such a way that when more than one bio-material is to be tested, multiple digesters can be arranged in a row to carry out different experiments in each.

The Nano 45 model Biogas digesters are also useful for diverse educational institutions such as schools and universities. It will be more convenient to use Nano 45 biogas plants to explain the component parts of a biogas digester to each student in each classroom in educational institutions. The Nano 45 biogas plant has been designed by incorporating all the components of a floating dome model biogas plant. By observing this directly, students will be able to gain first-hand understanding of the key components of biogas plants, the working of each component and their requirements. Such practical training will also help the students develop a proper understanding of the biogas project in the future.

As the awareness created among students during their school time spreads from them to their homes and from their homes to the community, the future generation will have a better awareness about biogas projects. There is no doubt that the Nano 45 Biogas digester can become a part of the learning process in educational institutions, and thereby used as a most convenient learning tool by government institutions as well as non-governmental organizations working in the field of awareness. As, such live training programs help to create a high level of awareness, this will enable every country across the nation to formulate and implement programs facilitating the availability of Nano 45 model Biogas plants to all educational institutions. Instead of teaching a technology several times, practical training given once will create an in-depth knowledge in students. Considering these factors, it can be seen that the Nano 45 model biogas plants can play a crucial role in the field of education.

Digesters for Demonstration

Exhibitions are as important as research and development programs in the spread of technology. Live demonstrations can play a significant role in raising public awareness on an advanced technology. It was in such a situation that digesters suitable for exhibitions became necessary to raise the awareness of the public about the production of biogas from bio-waste. If larger sized biogas digesters are used instead of Nano 45 model biogas plants for live demonstrations, more biogas can be produced and exhibitions can be conducted for longer periods and bigger burners can also be used.

Nano 160 model digesters have been designed such that it is suitable for such exhibitions. This biogas plant has a digester of capacity 100 litres and a gas collector having storage capacity of 60 litre. Their digester and gas collector is also made from fiberglass. Through this their long term performance can be ensured. Unlike Nano 45 model biogas plants, a single burner stove is also available along with Nano 160 model plants.

Chapter 11- Pic:2 Nano 160 Digester

As the biogas production from Nano160 model is more, training programs and exhibitions using single burner stoves can be carried out more efficiently. As these are bigger in size, their weight will also be comparatively higher.

A Nano 45 biogas plants can process a maximum of 50 gram of food waste per day, Nano160 model plants can process up to 150 gram of food waste. Depending on the structure of the raw material deposited for treatment, these digesters can produce up to 50 litre of biogas per day. Depending on the atmospheric temperature of the place where the digester is installed, there is a possibility of variation in the quantity of the biogas produced. A special type of trolley stand has also been designed for shifting Nano160 plants from one place to another. Such accessories can be selected by the beneficiaries as per their requirement along with these Nano digesters.

There are often situations where temporarily functional plants need to be provided to institutions engaged in the work of setting up biogas plants, to conduct operational training for their beneficiaries and to exhibit in public places. Nano160 model biogas plants have been designed in a way that is most suitable to fulfill such needs. Local self-governments and other local government institutions that are just starting to implement a new biogas project often request to install exhibition plants at the initial stage. Such exhibition plants are not required after a certain time. Nano160 model biogas plants are the most convenient for such temporary purposes.

In order to get high performance from Nano biogas plants, it is more advisable to grind the bio-wastes deposited in these into a paste and then pour it into the digester. When the biogas produced in a day is fully utilized, the gas collector is seen to fill up with biogas the next day itself. This will enable the biogas stove to function for a certain period of time every day. If the digester is not in continuous use for a long period after it has been used for a fixed interval of time, the digester can be kept in a safe place after all the bio-wastes inside it is fully drained out and the inside of the digester is washed and cleaned thoroughly so that there is no water content inside the digester. When needed, it can be re-activated and used further. To start reusing the plant, the steps followed while activating the plant for the first time need to be repeated.

Apart from the single burner stove which is available along with the Nano160 digester, experimental burners can also be obtained, if required. When experimental burners are used in Nano160 model biogas plants, the biogas produced will be sufficient to operate burners for a longer time when compared to stoves. Experimental burners will be more suitable in situations where exhibitions have to be conducted for a long period of time. If an additional experimental burner is also purchased along with the Nano160 biogas plant, biogas stove and experimental burner can be used alternately as and when required.

Chapter 11- Pic:3 Nano plant with stove

As exhibitions on Nano45 and Nano160 models spread through social media, it is seen that many people start demanding for such Nano plants for meeting their domestic needs. Such people do not pay any heed to the fact that only a very small quantity of biogas is generated from such models.

Nano biogas digesters have been designed exclusively for research and exhibition purposes alone. Many people often get mislead like this when they see the advertisements on such models. In order to ensure the availability of Nano biogas plants, they have been designed in such a way that they can be produced on an industrial scale. When these are produced on an industrial basis, construction using fiberglass will not be practical. This is because; only one or two plants can be produced from a mould in a day. Due to this, many moulds may be required in situations where large numbers of plants are required. This often requires a huge investment.

When Nano biogas plants are constructed for experimental needs on an industrial basis, blow molding technology will be most suitable. The metal parts inside the digester can be made either from stainless steel or powder coated iron. When the outer covering of the digester is constructed using blow molding technology, hundreds of digesters and gas collectors can be built daily. To manufacture a product by blow molding, approximately 1 to 3 minutes is only required.

The feature of fiberglass product manufacturing is that the construction cost of fiberglass moulds is comparatively low. Even if one mould gets damaged, a new mould can be made within a short period of time. Comparatively, only a low investment is required to establish fiberglass product manufacturing units. But for establishing an industrial unit which manufactures products by blow molding, a large investment is required. In blow molding, manufacturing moulds for each product requires a huge amount of money. Another feature that is worth mentioning is that the money invested for this is regained within a short period of time as the production takes place on a large-scale.

In the early days, Gober gas plants which can be used by cattle owners and large sized plants used for industrial purposes were the only plants that were popular. But as a result of technological advancement, different types of biogas plants that can be utilized by all sections of the population, including Nano biogas plants became available. There is no doubt that the availability of such micro digesters will accelerate the research activities in this field. It is an undoubted fact that Nano Biogas plants will become an integral part of all the institutions engaged in research related to bio-materials.

Nano Biogas Plants for Domestic use

As the number of residential complexes increased with the rise in population, the quantity of waste produced daily also increased considerably. In many places, the timely treatment of waste generated on a daily basis remains a major issue. The current trend of limited number of members in families residing in urban areas is also increasing. In many of the households only aged people are present. The food habits of the people in many households are also changing. The tendency to reduce one's own cooking and rely on online food delivery systems is also increasing day by day. All these also leads to a reduction in the quantity of food waste or kitchen waste generated from households. Even then approximately 500 gm to 1kg of bio-waste is generated on a daily basis from such households. Even if people living in households, especially flats and apartments in cities wish to install biogas plants for waste treatment, are unable to do so due to space constraints. In the light of the demand from many areas that it would be much more convenient if a suitable biogas plant is made available for such people, necessary steps were taken by BIOTECH to design Nano plants suitable for this and as a result, two plants of the same size but of different diameters were designed. They are Nano 500V1 and Nano 500V2.

Chapter 11- Pic:4 Nano 500 V1

The daily treatment capacity of such plants is 500 gm of bio-waste. But under special circumstances this can be raised up to 1kg. Nano 500 V1 has a diameter of 80 cm and height of 95.cm. The Nano 500 V2 model has been designed by reducing the diameter and increasing the height in such a way that it is suitable for those with very little space. Nano 500 model biogas plant is most suitable for those who are interested in producing biogas from a very small quantity of food waste. The most prominent feature of Nano500 model biogas plants is that they can be installed in flats and apartments.

Nano biogas plants, which can be installed in a very short time (approximately 1 hour) have been designed in such a way that they can be installed on the balcony of a house or on the floor outside the house or on the terrace above the house. The digester of Nano 500 V1 model has a storage capacity of 350 liter and gas collector of 150liter storage capacity. The total volume of a Nano500 V1 model biogas plant is 500 liter including the digester and gas collector. The Nano 500 V2 model biogas plant has been designed in such a way that its digester has a storage capacity of 360 liters and the gas collector has a gas storage capacity of 140 liters. Nano biogas plants are most suitable for making the idea of "One biogas digester for every home" a reality. They have been designed in a way that is most suitable for the timely treatment of household wastes for those who receive only small quantities of bio-waste.

Chapter 11- Pic:5 Nano 500 V2 for domestic Use

With each Nano biogas plant, a 5 meter gas pipe, one single burner stove, gas control valve set to be attached to the biogas collector, additional weight to be mounted on top of the gas collector, fiberglass cover for the weight, inlet tank for depositing waste into the digester and cover for keeping the inlet tank closed are available as standard fittings. A detailed user manual containing all the necessary guidelines for putting together the various parts of the plant, initiating the plant and for operating the plant daily is also available along with each digester. Once the user manual containing simple instructions which can easily be understood by any beneficiary is received, anyone can activate the Nano Biogas plants and start using it for their day to day activities without any external help. Bio-culture which is required for activating domestic nano biogas plants initially will also be provided along with these plants for those in need. Biogas production is seen to start within a very short time in biogas digesters which are activated using bio-culture mechanism.

Both V1 and V2 plants of the Nano 500 model are designed for large-scale production through blow molding. In situations where domestic biogas plants are required on a large-scale in a country, it is possible to produce one plant every three minutes through blow molding. For this, separate moulds will be needed for the digester and gas collector.

The important factors to consider when selecting a digester for biogas production are their size, the space required for installing the plant and their biogas storage capacity. The quantity and type of waste to be treated each day is also an important factor to be considered. If a Nano 500 model biogas plant is installed in a house, and if waste more than its treatment capacity is deposited, the waste thus deposited in the digester will come out of it before it is completely processed. It has been found that this gradually reduces the efficiency of the plant and makes the plant nonfunctional within a very short period of time.

Nano 500 model biogas plants are most suitable for homes where bio-waste is available only in small quantities. If people who have large quantities of bio waste and kitchen waste install suitable sized plants after examining in detail the amount of waste generated each day, such plants will function efficiently for a long time without any significant damages. Suitable sized plants should be recommended to a beneficiary only on the basis of the quantity of bio waste available per day from houses and institutions.

If plants of sufficient sizes are installed instead of small sized plants for the treatment of large quantities of waste, high quantities of biogas and organic manure can be obtained from such plants. In addition, it also helps to get a quick return on investment.

There is no doubt that the Nano 500 model biogas plants are the most suitable household appliance for domestic purposes. Considering the fact that every appliance has its own advantages and limitations, if an appliance is used solely for the purpose for which it has been designed, the proper benefit can be obtained. Whatsoever, Nano biogas plants do not have any significant functional defects. As these are made from quality fiberglass, they will also have a long service life. A Nano500 biogas plant has been designed in such a way that it can function efficiently for more than 15 years.

Nano biogas plants do not experience any significant functional impairments related to day-to-day operations. If the bio-waste deposited in the plant is of large size, they can be ground into a paste using a mini crusher before depositing into the digester to prevent the bio-waste from accumulating in the inlet pipe and clogging the pipe.

If the condensed water formed in the gas pipe is removed at regular intervals, the obstructions that may arise in the gas line when biogas reaches the stove can also be avoided. Fuel efficiency can be maintained by keeping the biogas burner clean. Frequent use of gas control valves attached to the stove burner can cause accumulation of soot, grease etc which may interrupt the smooth operation of these valves. This can be rectified by cleaning the valves at regular intervals. Instructions for this will be given in the user manual provided when setting up the plant. If floating model Nano digesters become non-functional due to the accumulation of bio-wastes inside the digester, the accumulated treated waste can easily be removed from the digester by lifting the gas collector out from the digester and re-attaching it to the digester after the waste is removed. This can easily be done by the beneficiaries themselves. If the bio-wastes are passed into the digester only after crushing it into a paste, the situation of the inlet pipe getting blocked can be avoided to a large extent.

The use of slurry from Domestic Nano plants

When Nano biogas plants are installed for domestic purposes, the use of slurry coming out after treatment becomes a problem. People residing within very limited spaces often pour the slurry out into the drainage or open spaces as they do not have a proper understanding about the usage of slurry. But not many people know that the slurry coming out after treatment from the biogas plants is a rich organic manure. It has already been proven that biogas slurry can be collected and used as bio-manure for agricultural purposes.

If in every house installing Nano biogas plants, vegetable cultivation is also done in five or six flower pots along with the plant, not only can the biogas slurry be used properly but also good quality and reliable vegetables can be obtained.

The slurry obtained each day from the plant can be mixed with one or two times of water and sprinkled over the leaves of the vegetables or poured gently at the base of the plant. Those who do not have the facility to do vegetable cultivation can pour biogas slurry into the base of other plants also. The slurry coming out from biogas plants which are loaded with bio-wastes more than the treatment capacity of the plant every day, will have a foul smell and high acid content. If such slurry is poured at the base of the plants, they are likely to get burnt. This is due to the high acid content in the slurry. In such conditions, the slurry must be diluted with water; five or six times the quantity of slurry before pouring it at the base of the plants.

There is no doubt that Nano biogas plants can play a significant role in the promotion and expansion of the domestic biogas project. This will also enable the benefit of the biogas project to reach all sections of the society within a short span of time. Nano Biogas plants are more than just a tool for biogas production but have the capability to attain multi-faceted benefits such as bio-waste disposal and production of bio manure. When people belonging to all sections of the society become beneficiaries of this project considering the fact that this technology is adequate for this purpose, it will help to bring about changes not only at the household level but also at the social and national level.

Recognizing the importance of this project, people working in this field should prepare suitable blueprints and guidelines and make this project a part of a popular movement. There is no doubt that the domestic Nano biogas project will become a popular movement if the government and other non-governmental organizations that spend crores of rupees for various developmental activities give due consideration to this project. The benefits earned from this by each household are numerous. In addition, all those working in this field need to be aware that this would be helpful not only to local self-government institutions in each area and the society but also helpful for conservationists in providing environmental protection.

Chapter 12

Socio - Economic Benefits From Using Biogas Plants

There are numerous financial benefits for private households, individuals, institutions and government institutions setting up biogas plants. Since the daily bio-wastes from households and institutions having biogas plants are getting treated at their source itself, the amount that had to be given to self-governing bodies or agencies for the collection and treatment of waste can be saved. With the timely disposal of bio-waste generated in public places with the help of biogas plants installed in public places, it is possible for the local self-government bodies to save the money spent annually for the collection and treatment of waste in public places.

The social benefits of the biogas project are numerous. Biogas plants play a major role in raising the living standards of the people. As the bio-wastes disposed of daily in public places are prevented from accumulating through timely treatment, it has become a permanent solution to the tendency of waste getting accumulated in public places. With the disappearance of waste heaps from public areas, the nuisance of stray dogs, flies, rodents such as rats can also be controlled.

By treating the waste received on a daily basis, it is possible to completely eliminate air pollution from the accumulated waste and create a healthier environment.

Biogas, a cheap, efficient and smokeless source of energy, produced by this process can be used for cooking, lighting, running different types of engines and also for the generation of electricity.

High-grade solid and liquid organic manure produced through bio-methanation technology can supplement and substitute the use of chemical fertilizers.

Extensive use of biogas plants will help to dispose of waste heaps in public places and prevent the spread of foul smelling wastewater from such waste heaps into the surrounding areas, thereby preventing the spread of flies, mosquitoes and a large number of germs and help in creating a hygienic environment.

The use of biogas as a cooking fuel helps to control the use of firewood and consequently to reduce deforestation and improve the level of environmental protection. The use of Methane, a key component of biogas, as a non-conventional energy source can greatly reduce the use of fossil fuels.

The social benefits from biogas plants are numerous and are beyond words as they can be tailored to the needs of each region.

Climate change and green house gas emissions can also be reduced by capturing methane through the installation of biogas plants.

Benefits from using biogas as a cooking fuel

The biogas produced through biomethanation technology is an efficient and effective fuel which can be used as an alternative to firewood, kerosene, cow dung cake and LPG. By using such a gas that does not produce smoke or soot will also help in improving the health of people, especially housemaids.

LPG is the main cooking fuel in most urban and rural households. The daily rise in the price of this fuel is now a matter of great concern to the general public. It is possible to produce biogas on a large scale by setting up biogas plants that are suitable for their needs in every household and small and large enterprise. The widespread use of biogas is conducive to various benefits.

The economic benefits of using biogas as a cooking fuel are numerous. Most importantly, LPG consumption can be reduced by more than 50 percent. It should also be noted that the use of other cooking fuels can be significantly reduced. More than 50 percentage of the amount spent per month for LPG can be saved. The cost of replacing the LPG cylinder each time can also be saved by more than 50 percent. The amount given for the disposal of bio-waste received on a daily basis can be significantly reduced. As the kitchen and utensils do not get dirty with soot, the money that has to be spent on buying cleaning materials can be saved.

Chapter 12- Pic:1 Use of biogas as cooking fuel

Chapter 12- Pic:2 A single burner biogas stove

Biogas is safer and less dangerous, compared to LPG and other similar fuels used for cooking. The use of biogas plants is very economical as there are no recurring expenses. In all public institutions like hospitals, schools, etc., the biogas produced from biogas plants can effectively be utilized for cooking, boiling water and for various other fuel purposes.

Each house will function as a mini power house when biogas plants become widely used in all households. The womenfolk can also be saved from various respiratory illnesses such as asthma which arises from the use of fuel wood to a certain extent.

The social benefits of using biogas as cooking fuel are also numerous. The use of firewood is significantly reduced when cooking with biogas in rural areas. This helps prevent deforestation for obtaining firewood. The reduction in tree destruction protects many trees and makes the air in the area cleaner. The money and time spent in searching for other cooking fuels can be saved.

The use of biogas also helps in escaping from various eye diseases caused from continuous exposure to smoke. Thus the use of biogas plants paves the way for improving the health and hygiene of all categories of people.

Role of Biogas project in different categories of people in the community

There are mainly three categories of people in the society, namely low income group, middle income group and high income group. Biogas plants can play a very important role in improving the life of all the three categories of people.

For those who belong to the low income group, biogas plants can be considered as a means for obtaining cooking gas and bio- manure at low costs and also a means for maintaining cleanliness through timely disposal of waste.

Those who belong to the middle income group can utilize the biogas plant as an instrument to dispose of the bio waste generated on a daily basis without causing any atmospheric pollution and also as an instrument for production of bio manure suitable for vegetable cultivation. For those belonging to the upper income group, biogas plants can be utilized as an instrument for timely waste management, by the house owners themselves without anybody's help. The biogas obtained can also be used as fuel for cooking.

By utilizing biogas as a cooking fuel substitute to LPG, biogas plants also becomes a means for reducing the use of LPG.

In slum areas, by the installation of decentralized biogas plants the problems of waste management and sanitation can be solved.

In coastal and slum areas lacking proper sanitation facilities, the installation of toilet-linked biogas plants along with the densely constructed houses in these areas will help to solve the sanitation problems to a certain extent.

It has been proved from several studies that installation of toilet linked biogas plants in large numbers will help in the treatment of bio- wastes, increase the hygienic condition of the society and produce cooking gas at a low cost, at the same time.

Domestic biogas plants are best suited for timely waste management for those who belong to the economically advanced category. The cooking fuel used by such people will naturally be LPG. They can also save more than 50 percent of LPG by using biogas for cooking. Growing their own organic vegetables is something that is of interest to everyone now. Slurry which is obtained from biogas plant can be used as a rich organic manure for organic vegetable cultivation in the kitchen garden.

The bio-waste treatment biogas project aims to find a permanent solution to the diverse problems that people from each section of the society face. The time has come for national and international public institutions, private development agencies and especially government agencies to come forward to give the necessary publicity and appropriate incentives to those who install biogas plants.

Benefits to the villagers from using biogas plants

With the establishment of biogas plants, the villagers can permanently eliminate the problem of smoke and soot emitted from the traditional firewood chulhas used in their kitchens. This will therefore help in improving the cleanliness of the houses and also helps in controlling the use of fuels such as firewood, cow dung cake, kerosene etc.

Rural women can also get relief from eye and respiratory diseases with the elimination of soot and smoke. Villagers can also keep their houses and surroundings clean by eliminating flies and mosquitoes that breed on the waste dumped with the help of biogas plants. As the treated slurry discharged from the plant does not emit any bad odour, the house and surroundings can be kept clean and free from bad odours.

Those villagers who attach their toilets to the biogas plants can produce more biogas and organic manure which can be used for agricultural purposes. Mixing the treated slurry with other organic refuses will help to quicken the decomposition of waste into enriched compost.

The expansion of the biogas project in rural areas will lead to many benefits. It directly and indirectly contributes to the overall upliftment of the rural population by increasing the income of the rural population, protecting their health and creating new employment opportunities. The economic benefits of using biogas can be considered as an indirect source of daily income. There are many machines that use biogas as fuel. The establishment of small scale industries that can install and operate such machines will also help in the industrial growth of the rural areas. Unemployed people in rural areas have the opportunity to work in a number of innovative areas, such as manufacturing biogas plants, setting up biogas plants for beneficiaries and repairing of plants.

Agriculture is one of the main occupations of the rural people. The slurry obtained after processing from a biogas plant can be used as a rich organic manure for agriculture. This helps to significantly reduce the use of chemical fertilizers. The quality of the soil and the ability and tendency of the soil to store water is also seen to increase. Biogas slurry is a good quality pesticide. Biogas slurry can be used to prevent many types of small insects from attacking plants. The ability of biogas slurry to keep pests away from vegetables and other plants is noteworthy.

Organic Manure

The organic material discharged from biogas plants after treatment can be used as organic manure. This is very good for the growth of all types of plants. This can be used as fertilizer and applied to the plants directly or after mixing it with water. The solid manure can be separated from the bio-slurry, if necessary, through special filters.

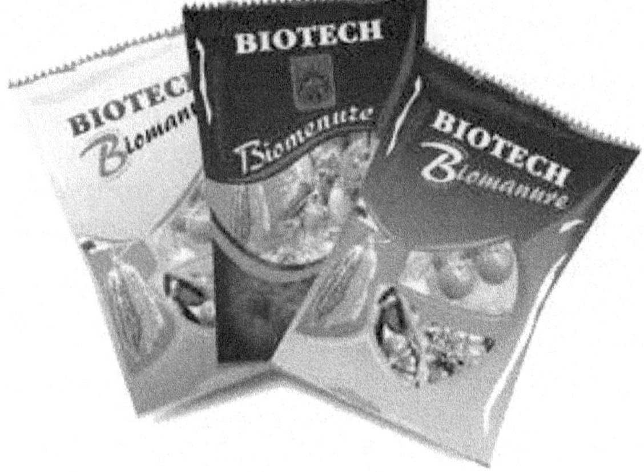

Chapter 12- Pic:3 Anaerobically treated solid manure ready for sale

Filtered liquid can be utilized immediately and the solid manure can be kept for later use. Treated solid manure can also be collected directly from pre-digesters specially designed for organic waste treatment. The bio manure produced through bio waste treatment is a better substitute for chemical fertilizer. Through the utilisation of this bio manure, huge amounts of money spent for purchasing chemical fertilizers can be saved to a great extent. The growth ratio of vegetables and plants will improve within a short period through the use of the bio- liquid fertilizer. The resistance of plants to insects will also get improved. The water storage capacity of the soil will be improved through the application of solid bio manure. This will also help in keeping out insects from to a certain extent.

Two types of byproducts are produced when bio-materials are disposed of in biogas plants, namely solid bio-fertilizer and liquid bio-fertilizer. Depending on the composition of the raw materials used in domestic biogas plants, the composition of the fertilizer obtained will also vary. The slurry which comes out of biogas plants treating cattle dung, other animal and bird droppings and human excreta will be in the form a paste. This will be a mixture of solid and liquid fertilizers.

The slurry which comes out after treatment from biogas plants treating kitchen waste will be in liquid form. Solid waste materials having an excess of fiber content will remain inside the digester after treatment. This can be taken out by opening the digester once in two or three years and used as bio-fertilizer.

In plants having pre-digesters attached along with their digesters, the solid bio-fertilizer which comes out after treatment from the plant can be taken out of the pre-digester at regular intervals. In industrial biogas plants the processed bio-fertilizer comes out in the form of slurry. The liquid obtained after removing the water content from this can be collected as liquid bio-fertilizer and the solid bio-fertilizer and both of these can be marketed separately.

Even if cattle dung is treated in large biogas plants installed on an industrial basis, slurry de-watering machines are used to collect the solid organic manure. Timely removal of the processed slurry that comes out from biogas plants on farms is of utmost importance. Daily tons of cattle dung comes out. Timely treatment of these is most essential to maintain a healthy environment on the farm. As the amount of slurry that comes daily from the plant is almost equal to the amount of waste that is treated, their timely treatment is most essential.

Liquid manure

Organic waste comes out after treatment from bio-waste treatment biogas plants in two forms. They will be available in the form of solid and liquid bio-fertilizer. Liquid fertilizer discharged from the plant after treatment can be mixed with water and directly applied to the plants.

Chapter 12- Pic:4 Anaerobically treated liquid manure ready for sale

In the case of cow dung based gober gas plants, treated materials are discharged in semi-solid form. This can be used as a fertilizer for plants and can also be used for the production of compost or earthworm compost by mixing it with agro waste.

Chapter 13

Carbon Credit Project

Air Pollution

Millions of years have passed since the earth became habitable. Agriculture and allied industries have been practiced in each region since ancient times based on the local climate. Each year, there were different weather conditions such as rain, snow and summer at regular intervals. Different crops were cultivated in each region according to the local climate.

Gradual population growth and industrialization has become the cause for man's desire for a higher standard of living. The day to day growth in technology also accelerated the growth of industries. With the increase in population, the number of residential complexes also increased day by day. As the next step to destroying forest lands for constructing residential complexes, man has reached the stage of constructing buildings even in the sea apart from the buildings already constructed on water bodies.

With the gradual rise of industrial growth, the effluents released from industries also gradually increased. Pollution from industries began to reflect at various levels. This began causing air pollution, environmental pollution and water pollution and thereby adversely affecting all sectors. Until recently, no one cared about the fact that pollution in all spheres, such as land, air and water was beginning to take its toll on human beings.

Chapter 13- Pic:1 Air pollution from factories

With the increase in population along with the industrial growth, the level of emissions from households and institutions also began to increase exponentially. Only a handful of developed countries have succeeded in properly disposing of their daily waste in a timely manner. In all other countries, waste began to accumulate due to the inability to adopt proper waste management practices. This led to a resurgence of serious pollution levels. While the wastewater from waste heaps contaminated the freshwater sources, the gases emitted from the wastes caused air pollution. The resulting pollution was such that it was impossible for people to live in the area where the waste heaps were located.

Thus the pollution which increased from different sources day by day gradually began to adversely affect the habitat of this beautiful earth. Things got to the point where many species became extinct and many others became endangered. Daily activities and unscientific approaches kept increasing the level of pollution again. As atmospheric pollution began to violate all its boundaries, its result could be seen all around. Gradual changes in nature have brought things beyond what we can bear.

Climate Change

Due to the gradual change in the climate, many areas which used to be cultivated according to the climatic conditions could not continue their cultivation. The main reason for this was the onset of severe summers during the rainy season, severe droughts during the winters and torrential rains during the hot summer months. This is the phenomenon of climate change.

Chapter 13- Pic:2 Unexpected Flood

The main cause of climate change is global warming. Atmospheric warming to a certain extent and the problems caused by it are beginning to seriously affect the habitats of humans and other living things on earth. As global warming began to affect the oceans, the seawater began to heat up rapidly. As such changes started occurring in the seawater also, hurricanes which happened only once or twice a year in ancient times gradually became a regular occurrence. Sudden torrential downpours with the increase in sea water evaporation led to massive damages and deaths. Frequent low pressures also caused continuous hurricanes and tornadoes known as cloudbursts.

Chapter 13- Pic:3 Unexpected Flood

If global warming continues to rise as it is today, many low-lying areas near the oceans will be flooded in the near future, according to reports released by the meteorological centers. News about the need to control climate change and the global warming that causes it comes in daily. If global warming is to be controlled, there is an urgent need to bring about control over the things that are causing it and man-made causes. Air pollution from industrial, residential, public and private institutions can be significantly reduced by treating the waste that comes out in a timely manner. Biogas production or biogas collection from these is the most suitable and cost-effective technology for timely treatment of bio-waste. These matters have been discussed to understand how important the role of biogas plants is in minimizing the harsh effects of climate change that the world is facing today. Recognizing the importance and the need for biogas plants, necessary steps to establish maximum biogas plants in all areas suitable for their effective operation needs to be completed urgently.

Methane (CH_4), which is the main component of biogas belongs to the green house gas category. Different studies have revealed that methane is 22 times more dangerous than carbon dioxide. Therefore it is very essential to prevent the spread of methane into the atmosphere.

The waste dumped in various locations by the public generates huge quantities of methane which get emitted into the atmosphere. Similarly, collecting and retaining the waste in public places for long periods will also lead to the spread of methane into the atmosphere. Biogas plants play an important role in preventing the spreading of methane. Due to this, the biogas project is also in known as methane capturing project.

Carbon Credit

From a 1 m3 sized biogas plant, we can generate 365 m3 of biogas in a year at the rate of 1 m3 per day. This is equal to 3.5 metric tons of carbon dioxide. Thus the installation of a 1 m3 sized biogas plant prevents the emission of 3.5 metric tons of carbon dioxide into the atmosphere. The prevention of the spread of one metric ton of carbon dioxide into the atmosphere is considered as one carbon credit. Thus 3.5 carbon credits can be earned annually by the installation of a 1 m3 size biogas plant.

Carbon credit is an area with huge marketing potential globally. Industries and institutions that are unable to implement their own pollution control systems tend to purchase pollution control schemes that are commensurate with the amount of pollution they produce. It is in this context, that the carbon credit project becomes important. But carbon credit cannot be earned by installing a single biogas plant of 1 m3 size. This is because; the registration for carbon credit and the processing of the project at different stages are very expensive.

For this reason, biogas plants installed in different places can be bundled together as one project and carbon credit registration can be obtained. This is known as project bundling. On registration of projects as bundles, the members of that group can earn regular income from the sale of carbon credit from their plants.

Tens of thousands of domestic, Institutional and Industrial biogas plants are set up each year by government agencies and various other agencies in each region, such plants can be bundled together locally to obtain carbon credit registration. On completion of the registration procedures, the carbon credits can be sold in special markets meant for this. The fact that one can make financial profit by selling carbon credits in such markets is also one of the future prospects of the biogas project.

Chapter 14

Future Possibilities of Bio Gas Plants

As in all other fields, tremendous technological advancements are taking place in the biogas sector also. The time is not far away when biogas plants will become an indispensible tool in the daily life of humans. The proliferation of biogas plants has already proven to be equally beneficial to all sections of the society. The need of the hour is for all sections of the society to start using biogas plants as part of their social, economic and cultural growth. The future potential of the biogas project is immense as biogas plants can be utilized with equal importance for personal gains such as hygiene, health care, energy needs, social upliftment and ensuring income, which are urgently needed in daily life and in social sectors such as increasing job opportunities and prevention of environmental pollution.

Environmental pollution can be controlled to a large extent by setting up biogas plants as a means for timely disposal of bio-waste in technologically growing and technologically backward areas. This will also help in maintaining adequate hygiene in each area. By being able to dispose of the bio-waste generated on a daily basis from individual households and institutions in the respective biogas plants in each area, there will also be a change in the situation where waste gets accumulated in public places.

Timely disposal of wastes prevents the spread of parasites that grow on them, thus reducing the incidence of diseases and various infectious diseases caused by rodents such as rats, flies, mosquitoes and stray dogs and also help in maintaining the health of the population. The future expansion of biogas plants will further improve the health of the people as the use of biogas as fuel will save the housewives from the problems caused by soot and smoke.

Energy Sector

In the future, biogas plants will play an important role in the energy sector. The use of biogas as a cooking fuel can also lead to a significant reduction in the use of other fuels. Biogas plants can also improve the economy of the people by saving the money which they had to spend on alternative fuels. Since biogas is available free of cost every day, no matter how high the price of other fuels is, it does not affect the biogas production. With the proliferation of biogas plants in the domestic sector, biogas plants can have a significant impact on the domestic sector in future as it becomes an additional source of income for every household.

The establishment of biogas plants will enable private establishments and public institutions working under various sectors to reap all the above said benefits.

Industrial biogas plants can also play a significant role in social, economic and technological development in the future. By setting up industrial biogas plants along national highways and other areas, these plants can be used for waste treatment in the surrounding areas. The biogas produced in industrial biogas plants can be purified and used as vehicle fuel. High quality bio CNG filling pumps can also be installed in conjunction with industrial plants along the national highways.

Research and development work on innovative pollution reduction projects is progressing at an accelerated pace around the world. As a part of this, huge changes are taking place in the automotive industry as well. There is also a situation where vehicle manufacturers are turning to the research and development of electric vehicles. As electric vehicles become more widespread, their charging stations will also become as widespread as the petrol pumps we now see on public roads. It is also possible to supply the required power to the charging stations by generating electricity from the biogas produced in industrial biogas plants installed along roadsides. It will be an eco-friendly project as it is green electricity.

In the future, biogas plants will be of high demand in the field of electric vehicle charging. This will further brighten the future prospects of industrial biogas plants. As portable biogas plants become an integral part of every household, many manufacturing units will be available for their production. It will also lead to the employment of many people. With the expansion of portable plants, many job opportunities will be created in the future, in the areas of their installation and repair. This has led to the entry of many individuals and institutions into the biogas sector on a short and long-term basis.

Domestic Sector

Attitudes towards biogas plants may change in the future. There may be a law in all areas that stipulates that when a building is constructed, a sewage treatment biogas plant must be built as well. A few local bodies in Kerala have already enacted the law stating that waste treatment biogas plants must be installed in order to obtain a license to construct buildings. Environmentalists see this as a bold decision that can set an example for others in the future.

Research for increasing the efficiency of biogas plants is underway in many parts of the world. Research is also in progress to accelerate biogas production, which is now a purely biological process. Many companies have already developed bio-cultures that accelerate biogas production. These cultures, which are available in liquid and powder form, are found to increase biogas production by dissolving a certain amount of it in water and pouring it along with the bio-waste into the digester.

In the future, there may be variations even in the models of domestic biogas plants. Nano biogas plants that can be placed under the kitchen sink in every household are also available. Biogas technology might evolve in such a way that the bio-wastes received each time is crushed into a paste with the help of a compact crusher and transported to the biogas plant under the sink. The bio-waste that reaches these sophisticated digesters will be processed in a very short time and biogas will be produced efficiently. The biogas storage methods of such innovative digesters may also change. If fuel cells that get charged as soon as biogas is produced are attached in conjunction with biogas plants, the need for storing biogas can be eliminated. Such fuel cells that get charged can be used to meet electrical needs.

Fuel cells can play an important role in the future prospects of biogas plants. Fuel cells are being researched in different parts of the world. Already several fuel cells in different models have been developed. The high cost of fuel cells is the reason why they could not be made more popular yet. Research to develop less expensive fuel cells is also progressing in various parts of the world. Fuel cell is charged by breaking down methane which is the main component of biogas. With the proliferation of fuel cells, there will be no need for the kind of gas collectors currently available for biogas storage. This is possible as fuel cell gets charged as soon as biogas is produced.

The situation around the world is now conducive for the biogas project to expand beyond what we see today. There is no doubt that the biogas project is a promise of tomorrow that will play a crucial role in tackling the problems of growing fuel shortages and accumulation of waste.

Role of BIOTECH INDIA

Established in 1994, BIOTECH India Renewable Energy (BIOTECH) has a unique role in research, development and implementation of biogas projects. BIOTECH India has adopted a working method of interacting with more than 32,000 beneficiaries on a regular basis and providing timely assistance to them.

Our research department pays special attention to analyze carefully the feedback of BIOTECH beneficiaries and bring timely reforms in each project. We are implementing biogas projects at various levels in collaboration with government and non-government organizations.

Household biogas plants are being widely installed in coordination with local self-government bodies to realize home source waste treatment and cooking gas production. BIOTECH has developed biogas plants of various sizes that can produce biogas by treating organic waste and sewage from markets and abattoirs under the jurisdiction of local self-government bodies. Generation of electricity using biogas from Institutional level biogas plants has gained immense popularity.

BIOTECH is also imparting necessary training for those who want to enter the field by making the biogas project a career venture. Various training programs are available online. We are also running a website www.biogascourse.com for this purpose. Practical training related to the construction of biogas plants is also available at its training division.

BIOTECH has already developed several biogas plants of various types to achieve world-class performance in domestic biogas projects. Various types of digesters are available, such as floating dome type portable biogas plants. The Portable Water Jacket and Portable Ultra models are more popular. Turbo biogas plants are also highly efficient portable plants for obtaining biogas under high pressure without mechanical assistance from portable biogas plants. Turbo model biogas plants are designed to be suitable for domestic biogas plants to produce on an industrial basis. Those interested in manufacturing and marketing portable biogas plants are provided special training and license by BIOTECH.

A Tower garden that can be used with a biogas plant is best suited for processing slowly decomposing organic waste. The composter garden developed is designed to enhance the efficiency of domestic biogas plants. Our research arm has also contributed pre-digesters to control scum formation in domestic biogas plants using kitchen waste.

BIOTECH are also working as a technology partner of the National Institute of Rural Development and Panchayati Raj (NIRD&PR), under the Ministry of Rural Development, Government of India. BIOTECH has also set up a Biogas Resource Center inside the campus of NIRD&PR to provide training on various biogas projects.

By installing biogas plants in every household and every establishment, small and large, where space is available, cost-effective decentralized waste management can be made a reality. BIOTECH India is playing a crucial role in the future development of the biogas project. Our main objective is to bring the biogas project to all sections of the society. More details about BIOTECH INDIA is available from its web site www.biotech-india.org

Social media links of BIOTECH INDIA

* Use following QR Code for quick access *

Website
www.biotech-india.org

Instagram Id
biotech_india

Facebook Page
BIOTECH INDIA

Youtube Channel
BIOTECH Renewable Energy
youtube.com/@biotechhq

Dr. A. SAJIDAS
www.drsajidas.com

Index

Notes

Notes

www.ingramcontent.com/pod-product-compliance
Ingram Content Group UK Ltd.
Pitfield, Milton Keynes, MK11 3LW, UK
UKHW051307240426
12071UKWH00037B/964

9 789359 807416